Technology and Creativity

■

Technology
and
Creativity

□ □ □

Subrata Dasgupta

New York   Oxford
OXFORD UNIVERSITY PRESS
1996

Oxford University Press

Oxford   New York
Athens   Auckland   Bangkok   Bombay
Calcutta   Cape Town   Dar es Salaam   Delhi
Florence   Hong Kong   Istanbul   Karachi
Kuala Lumpur   Madras   Madrid   Melbourne
Mexico City   Nairobi   Paris   Singapore
Taipei   Tokyo   Toronto

and associated companies in
Berlin   Ibadan

Copyright © 1996 by Oxford University Press, Inc.

Published by Oxford University Press, Inc.,
198 Madison Avenue, New York, New York 10016

Library of Congress Cataloging-in-Publication Data
Dasgupta, Subrata.
Technology and creativity / Subrata Dasgupta.
p. cm. Includes bibliographical references and index.
ISBN 0-19-509688-6 (cloth)
1. Creative ability in technology.   2. Creative thinking.
I. Title.
T49.5.D39   1996
601'.9—dc20   95-20086

1 3 5 7 9 8 6 4 2

Printed in the United States of America
on acid-free paper

To
Biman Dasgupta
Jashodhara Bagchi
and
Malay Dasgupta
who helped shape the landscape
of *my* mind

Preface

In Plato's *Ion*, Socrates speaks of the making of poetry as a divine dispensation, a kind of madness sent down from heaven. His audience of one, the conceited "rhapsode" or reciter, Ion of Ephesus was thus pleased to be assured that when he recited the Homeric epics—a skill in which he claimed to be unequaled—he was possessed by, was a mouthpiece of, the gods. In the centuries that followed, many other luminaries—poets, composers, scientists and artists—would express similar sentiments about their own powers of creation.

Indeed, insofar as creativity has been the object of scrutiny, poetry, fiction, music, and science have all shared the limelight, as can be seen in Brewster Ghiselin's anthology of writings on *The Creative Process* (1952) or Rosamund Harding's *Anatomy of Inspiration* (1942). Yet these same works, immensely valuable though they are in what they contain, are as remarkable in what they exclude—notably, any discussion of, or writing on, *technological creativity*: the creativity entailed in the conception, invention, and design of original, useful artifacts.

Perhaps this boycott of technology from the realm of past discussions of creativity has to do, to some extent, with the fact that unlike the histories of literature, art, and science, the history of technology is largely one involving the unknown and the nameless, the humble and the unlettered. The technologist, moreover, whether ancient, medieval, or modern, is seen as a doer. That Rodin's *Thinker* could as much be the artisan or inventor as the philosopher or cosmologist, does not sit well with many people. Regrettably, inventors and engineers are themselves not exempt from such myopia: It is said that Archimedes, a mathematician, scientist, and engineer, held his practical work in such low esteem compared to his achievements in mathematics and mechanics, that he left no written record of his many inventions. More generally, the common reticence of inventors, designers, and engineers, their reluctance to reflect aloud on their craft—with a few exceptions—is, doubtless, further cause for why technology as a creative endeavor attracts far less attention than others.

There is yet another possible reason for this odd situation, especially in these modern times. And that is, technology is commonly viewed as the "mere" application of scientific principles to the solution of practical problems. History gives the lie to this perception. To begin with, if science is concerned with the *understanding* of nature, technology endeavors to *master* or *harness* it for practical ends. This distinction in aims has profound implications for how nature or natural phenomenon is perceived by science and by technology. In essence, the two approaches encapsulate what might be called *complementary* views. Furthermore, the practical necessity for mastery or control has far exceeded in urgency the purely intellectual and emotive desire to understand. Man had been making, treating, casting, and forging metals and alloys; constructing roads, bridges, dwellings, and public buildings; crafting boats and ships; and shaping the instruments and engines of war many thousands of years before the rational comprehension of their respective underlying scientific principles could even be contemplated.

Historically, technology is older than *Homo sapiens*. It reaches back to the hominids and the stone tools of the Lower Paleolithic Age about two and a half million years ago. Science, even in its earliest, most speculative form, is only a few thousand years old, and science as we know it began in about the fourteenth century. The mental process of inventing artifacts, thus, can scarcely be held to be an ancillary of the mental process of scientific discovery. Indeed, it is fair to claim that the earliest manifestation of creativity in humans and their immediate ancestors was in the realm of technology—much earlier, even, than the cave art we so admire, which is known to have been practiced twenty or twenty-five thousand years ago. Clearly, technology as a creative process deserves to be pondered in its own right, quite independent of science—although we might expect to see some points of contact, especially in the past three centuries.

The topic of this book is technology. Not its history—though history does play an important part—but its psychology: Technology as a creative process. It is a subject that is both fascinating and important. The fascination stems from the curiosity we all have about nature of creativity—in this case, about the nature of the inventing mind: What must the mind know in order that it may invent? How are technological ideas formed? What kind of thinking goes on in the mind when it translates human needs, aspirations, goals, and wants, often very abstract, vague, poorly conceived, barely a dream, into original and sometimes extraordinary forms of real world artifacts—buildings or bridges, machines, materials, engines, or algorithms—that must satisfy those same needs, wants, or goals in ways no other artifact had done before? We could, of course, like Plato's Socrates, ascribe the genius of invention to heavenly inspiration, but only if and when we are very sure that no earthly explanation is on offer. In this book, I have endeavored to combine the venerable practice of history with the brash young discipline of cognitive science in an effort to explore the terrain of the inventing mind—without having recourse to divine dispensation.

And of course, the subject is important because, regardless of our personal ideology concerning technology, regardless of whether each of us sees it as friend or foe, technology has forever been, is, and will remain, as inextricably entwined with human existence as the oxygen we breathe. To understand technological creativity is to enhance our understanding of technology as a human enterprise, and that is something we can ill afford to neglect.

This marriage of intellectual or scientific curiosity with practical, "real world" importance makes the topic of technological creativity irresistible. I invite the reader to join me in its exploration.

Lafayette, La.   S. D.
March 1995

Acknowledgments

This work was begun in Manchester, England, and completed in Lafayette, Louisiana, in the United States. I am greatly indebted to the endowments that have supported the two positions I have held in the course of writing this book—the Dowty Professorship in the University of Manchester Institute of Science and Technology (UMIST) and the Eminent Scholar Computer Science Chair in the University of Southwestern Louisiana (USL). They have given me the freedom and peace of mind so essential to the scholarly enterprise.

I must also take this opportunity to thank Ray Authement, President of USL, for his unfailing support, over many years, of my work, and for his indulgent tolerance of a philosophy of scientific scholarship that many would consider outside the mainstream or, worst still, downright eccentric.

I am extremely grateful to the following individuals: Claude Cech, who read the entire manuscript and responded with a detailed, thoughtful, and enormously cultured review; Donald Cardwell, whose comments on the manuscript from a historian's perspective were invaluable; Robert Weber, who read an early and partial version of the book and responded with constructive criticism of a level that befits someone who has written his own book on technological creativity; Walter Vincenti and Henry Petroski, who offered informed comments and discussions on matters pertaining to the contents that appear in chapters six and seven; Joe Marsh, who directed my attention to the literature on Benjamin Huntsman and who provided help on other bibliographical matters; and Terry Dartnall, who engaged in a memorable week of lively, sometimes pugnacious but always fruitful, debate on matters philosophical. Any errors or flaws that remain in the book are definitely not of their making. They are mine own, and, in the time-honored tradition, I can only murmur: mea culpa!

One of the great pleasures of working on a topic such as creativity is that it excites interest that cuts across disciplines or fields. I have had the pleasure and opportunity of giving lectures on this material to audiences that included philosophers, architects, computer scientists, civil, electrical, and mechanical engineers, psychologists, mathematicians, anthropologists, literary scholars, historians, and economists. The experience has been exhilarating. My special thanks to those who have given me the opportunity to expose the ideas discussed here in such eclectic gatherings—in particular, Amiya Bagchi, Terry Dartnall, John Gero, T. K. Ghoshal, Jeanette Parker, Tim Smithers, and Keith van Rijsbergen.

I thank Nancy Lastrapes Franks, who typed with good humor and patience the many versions of the manuscript, and Mark Radle who exercised his computational skill and imagination in producing the diagrams that appear here. I must also thank the editorial and production staff at Oxford University Press for their help throughout the editing and production process.

I thank the many publishers for granting me permission to quote excerpts or adapt figures from their copyrighted works, a list of which follows.

D. K. Simonton, *Scientific Genius: A Psychology of Science* © 1988 Cambridge University Press. Reprinted by permission of the publisher.

G. Sturt, *The Wheelwright's Shop* © 1993 Cambridge University Press. Reprinted by permission of the publisher.

H. W. Dickinson, *A Short History of the Steam Engine* © 1939 Cambridge University Press. Reprinted by permission of the publisher.

G. H. Hardy, *Ramanujam* © 1940 Cambridge University Press. Reprinted by permission of the publisher.

J. F. Baker, *The Steel Skeleton: Elastic Behaviour and Design* © 1954 Cambridge University Press. Reprinted by permission of the publisher.

E. Kant and A. Newell, "Problem Solving Techniques for the Design of Algorithms," *Information Processing and Management*, 20, 1–2, 97–118 © 1984. Reprinted with kind permission from Elsevier Science Limited, The Boulevard, Longford Lane, Kidlington OX5 1GB, U.K.

J. Gardner, *The Art of Fiction* © 1991 Vintage Books. Reprinted by permission of Alfred A. Knopf, Inc.

S. Smiles, *Industrial Biography: Iron Workers and Tool Makers* © 1967 David and Charles. Reprinted by permission of the publisher.

G. F. Watson, "Mesara Ibuka," *IEEE Spectrum* © 1991 IEEE. Reprinted by kind permission of the publisher.

R. K. Jurgen, "Jacob Rabinow," *IEEE Spectrum* © 1991 IEEE. Reprinted by kind permission of the publisher.

A. W. Burks and A. R. Burks, "The ENIAC: First General Purpose Electronic Computer," *Annals of the History of Computing* © 1981 AFIPS (now IEEE). Reprinted by kind permission of the publisher.

R. H. Thurston, *A History of the Growth of the Steam Engine* © 1939 Cornell University. Used by permission of the publisher, Cornell University Press.

E. W. Constant II, *Origins of the Turbojet Revolution* © 1980 Johns Hopkins University Press. Reprinted by kind permission of the publisher.

W. G. Vincenti, *What Engineers Know and How They Know It* © 1992 Johns Hopkins University Press. Reprinted by kind permission of the publisher.

G. Agricola, *De Re Metallica*, H. C. Hoover and L. H. Hoover (tr.) © 1950 Dover Publications. Reprinted by kind permission of the publisher.

Vitruvius, *Ten Books on Architecture* © 1960 Dover Publications. Reprinted by kind permission of the publisher.

A. P. Usher, *A History of Mechanical Inventions* © 1982 Miriam Usher Chrisman. Reprinted by kind permission of Dover Publications.

M. Polanyi, *Personal Knowledge* © 1962 University of Chicago Press. Reprinted by kind permission of the publisher.

E. T. Layton, Jr. "Mirror Image Twins: The Community of Science and Technology in 19th Century America," *Technology and Culture* © 1971 University of Chicago Press. Reprinted by kind permission of the publisher.

R. Kanigel, *The Man Who Knew Infinity* © 1991 Robert Kanigel. Reprinted with permission of Scribner, an imprint of Simon and Schuster, Inc.

A. W. Burks, "From ENIAC to the Stored Program Computer: Two Revolutions in Computers," in N. Metropolis, J. Howlett, and G-C. Rota (Eds.), *A History of Computing in the Twentieth Century* © 1980 Academic Press. Reprinted by kind permission of the publisher.

D. E. Knuth, "The Errors of TEX," *Software: Practice and Experience*, 19, 7, 607–685 © 1989 John Wiley and Sons. Reprinted by permission of John Wiley and Sons Ltd.

D. Pye, *The Nature and Aesthetics of Design* © 1978 The Herbert Press. Reprinted by kind permission of the publisher.

C. Alexander, *Notes on the Synthesis of Form*, Harvard University Press © 1964 the President and Fellows of Harvard College. Reprinted by permission of the publisher.

H. H. Goldstine, *The Computer from Pascal to von Neumann* © 1972 Princeton University Press. Reprinted by kind permission of the publisher.

W. Addis, *Structural Engineering: The Nature of Theory and Design* © 1990 Ellis-Horwood. Reprinted by kind permission of the publisher.

T. D. Crouch, "Why Wilbur and Orville? Some Thoughts on the Wright Brothers and the Process of Invention" in R. J. Weber and D. N. Perkins (eds.) *Inventive Minds.* © 1992 Oxford University Press, Inc. Reprinted by kind permission of the publisher.

Finally, a note of gratitude to my wife Sarmistha (Mithu) and our sons, Deep and Shome. As so often in the past, they suffered my preoccupations during the writing of this book with phlegmatic forbearance. This book is dedicated with great affection to three cousins who collectively yet in complementary ways exerted enormous and wonderful influence upon me during my teenage years.

Contents

1. **Prolegomenon, 3**

2. **Artifacts: Material and Abstract, 9**
   Material Artifacts, 10
   Inner and Outer Environments, 10
   Abstract Artifacts, 11
   A Distinction between Craft and Engineering, 12
   The Manufacturing Process as Artifact, 14
   The Janus-Faced Nature of Computer Software, 15
   The Fallibility of Artifacts, 17

3. **The Birth of Technological Problems, 20**
   Need, 20
   Dissatisfaction, 22
   Curiosity, 25
   Problems Become Goals, 27

4. **The Technologist as a Cognitive Agent, 29**
   Cognition at the Knowledge Level, 31
   Goals, 32
   Knowledge, 32
   Actions, 38
   Rational and Nonrational Actions, 42
   Bounded Rationality, 43
   Accessing Knowledge by Spreading Activation, 45
   The Dynamics of Knowledge Bodies, 46
   Knowledge Level Processes: Rational and Nonrational, 48
   Design as a Knowledge-Level Process, 49
   Radical or Inventive Design, 52

5. The Connection between Invention and Design, 53

The Test of Inventionhood, 55
Originality and Creativity, 55
The Criteria of Historical Originality, 58
The Act of "True" Invention, 64

6. Technology and Hypotheses, 66

The Hypothesis Law of Maturation, 66
Testability of the Hypothesis Law, 67
The Case of the First Superalloy for Gas-Turbine Blades, 69
The Case of the Convex-Hull Algorithm, 74
The Case of the Britannia Bridge, 78

7. The Process of Ideation, 87

The Classical View: Ineffability, 88
The Wallas Model, 88
Bisociation: The Combination of Unrelated Ideas, 91
The Darwinian Model: The Variation-Selection Theory, 92
Simonton's Model, 93
Vincenti's Model, 95
How Valid Is the Darwinian Model? 97
Superalloys, Revisited, 97
The Convex-Hull Algorithm, Revisited, 99
The Britannia Bridge, Revisited, 100
Ideation as a Knowledge-Level Process, 101
The Invention of Microprogramming as
   a Knowledge-Level Process, 103
Robert Stephenson's Conception of the Tubular Bridge, 110
The Role of Search in Ideation, 116
The Nature of Problem Search in the Design Process, 117

8. Creativity and the Evolution of Artifactual Forms, 122

The Newcomen Engine, 124
ENIAC: The First General-Purpose Electronic Computer, 134
Phylogeny Conditions Ontogeny, 146
Testability and Other Aspects of the Phylogeny Law, 146

9. The Nature of Technological Knowledge, 150

Scientific Knowledge in Technological Reasoning:
   The Reductionist View, 151
Technological Knowledge: Basic Science and
   Technological Theory, 152
Mathematical Knowledge in Technology, 157
Technological Knowledge: Operational Principles, 157
Stephenson's and Fairbairn's Knowledge, 158
Newcomen's Knowledge, Revisited, 160

Operational Principles in the Invention
    of Computing Devices, 162
Theory versus Operational Principles
    in Software Technology, 164
The Ontology of Operational Principles, 167
How Do Operational Principles Originate? 170
Operational Principles as Pictures, 177

**10. A Portrait of the Technologist as
    a Creative Being, 180**

The Technological Process Is Knowledge Rich, 180
The Technologist Is a Rational Being, 182
Ideation Is a Rational Process, 182
The Technologist Freely Espouses and Forms Hypotheses, 184
Technological Creativity Is Conditioned
    by Evolutionary History, 184
Epilogue, 185

Notes, 187

Bibliography, 215

Index, 225

Technology and Creativity

■

O! for a muse of fire, that will ascend
The brightest heaven of invention
—*Henry V*, Chorus 1

1

Prolegomenon

Anthropologists tell us that humans and their hominid predecessors have been conceiving, shaping, and using artifacts from as far back as the early Stone Ages. We now call this activity *technology*. Because of its ubiquity across time and space, the histories of cultures, societies, economies, and everyday life are inextricably entwined with the growth and evolution of technology.

In addition to its historical, social, and economic importance, technology has also a *psychology* and a *logic*. This connection, though not widely recognized, becomes quite obvious when we remember that technology is as much concerned with the *conceiving* of artifactual forms as with their actual making. Technology is, thus, a cognitive activity—it involves the use of knowledge and the faculties of reasoning, remembering, and understanding, as do all cognitive processes. It is this aspect of technology—as an intellectual, cognitive process—that is the focus and the concern of this book.

The idea that there is an intellectual dimension to technology is not particularly commonplace. The technologist is conventionally viewed as a doer—as a man or woman of action. In fact, the concept of the technologist as an "intellectual"—the idea that *Homo faber* is also *Homo cogitans*—may even cause some eyebrows to be raised among those who cling to the notion that the producing of material artifacts is anathema to the life of the mind. Technologists themselves are not entirely exempt from this belief—even those (or perhaps I should say especially those) responsible for the teaching of technology. I still recall vividly, even after the passage of three decades, my sense of dismay the day I began my undergraduate studies at one of India's oldest and most venerable engineering schools. In the lobby of the main building of this institution, I saw portraits of the nation's preeminent physical scientists. Most were fellows of the hallowed Royal Society, the world's oldest scientific body. But not a single engineer was on view!

The message could hardly have been more explicit: Technology (or engineering—the distinction is unimportant here) is merely the *application* of the relevant "basic" sciences to the making of artifacts; the intellectual meat of the sundry engineering disciplines lies in the underlying physical, chemical, and mathematical knowledge. Engineers though we aspired to be, we were to look for inspiration to Newton, Galileo and Lavoisier (or, their descendants such as those scientists whose pictures were on the wall) rather than James Watt, Thomas Edison, and Alexander Graham Bell.

3

This, then, seemed to be the message. And, in the course of the long, arduous, and otherwise substantial education I then went on to receive, nothing that my professors taught or said caused me to think the contrary. I now know that this message was simply a reflection of a powerful and long-held view of technology and its subservient relationship to science—a view which continues to be widely held even now; even, as I found recently, by so profound a thinker as the astrophysicist Subramaniam Chandrasekhar.[1]

But as I graduated from the "basic" sciences of the freshman year to the "engineering" sciences of the next—from physics and chemistry to strength of materials and applied thermodynamics—and these were then eventually supplanted by my metallurgical studies, I continued to ruminate sporadically on how such seemingly disparate elements of my education were *really* connected and what it was that we were *really* engaged in when we were "doing" engineering, and how "doing" engineering *really* differed from "doing" science.

Why, for instance, had the blast furnace taken the peculiar, gigantic satanic form it had and where was the "science" behind its shape? And how was it that the iron-making process had worked successfully long before its chemistry and its thermodynamics were understood? Most puzzling of all, how did the scientifically illiterate smiths in the shanty huts across the road from the campus forge red-hot steel so perfectly, knowing nothing about heat treatment or phase transformations in alloys? What knowledge did their skills embody? What did *they* know of metallurgy that we did not, schooled though we were in the most advanced theories of solid-state behavior?

Several years after obtaining my engineering degree (and having metamorphosed, in the meantime, into a computer scientist), I encountered Herbert Simon's 1969 book *The Sciences of the Artificial,* and light at last began to dawn.[2] I got my first glimpse of a possible resolution to my conundrum. And what an exhilarating glimpse it was! Simon's insight—that the conceiving and creation of artifacts necessitate a kind of reasoned thinking that goes beyond the natural sciences, that the world of the artificial contains its own logic which, though related to, is quite distinct from, the logic of the natural world—was a revelation. And as with most revelations it later seemed so obvious. Like Thomas Henry Huxley when he first learned of Darwin's theory, I wondered why what Simon said had not occurred to me before.

I now began to understand that the intellectual dimension of technology is only partially grounded in physics and chemistry. We can never comprehend technology on the basis of the natural sciences alone. For technology is concerned with the invention of artifactual forms—an activity that entails *human goals, aspirations* and *wants* and their satisfaction. The physical sciences, the "basic" or "fundamental" sciences, have nothing to say about goals or wants. Each technological discipline, then—each "science of the artificial," to use Simon's term—regardless of whether its domain of interest is alloys, bridges, cities, machines, satellites, software, or whatever, is basically *teleological* in nature. Any explanation of why an artifact has the form it does, why it was designed in such a way, cannot simply take the form of causal explanations based on physicochemical laws. It must necessarily be grounded in considerations of the purpose it is intended to serve *and* in the capabilities, limitations, and knowledge that were at hand in effecting that form. The practice of technology and of its understanding, thus, is governed as much by psychology as by physics.

Another aspect of Simon's arguments was the notion that while there are many distinct technological disciplines—civil, mechanical, and electrical engineering; the chemical, materials, and process technologies; agriculture, computer science, economic and social planning, architecture, and so on—one intellectually nontrivial activity lies common to all and is at the heart of each. This is the activity of *design*. Furthermore, it may be possible to investigate and elucidate the nature of design processes irrespective of the kind of artifact being designed. In other words, common to all of technology, at its intellectual root, is an artifact-independent discipline concerned with the nature of designing itself. Simon called this discipline the "science of design." Some modern researchers (including this writer) refer to it as *design theory*.[3]

I soon realized, after reading *Sciences*, that there were others who had also reflected upon the nature of design. For example, Christopher Jones and others had initiated the "design methods movement" of the 1960s; Christopher Alexander, a Berkeley architect, had sought to apply mathematical formalism to the design of an Indian village in his influential *Notes on the Synthesis of Form* (1964); and the industrial designer David Pye analyzed with elegance, clarity, and a great deal of originality some of the fundamental aspects of the design process in *The Nature of Design*, also published in 1964.[4] None of these writers, I might add, were from the traditional, mainstream engineering disciplines of my undergraduate days.

If the notion of technologist as intellectual is anathema for some, the idea of a *science* of design is as problematic for others. They would deny that there is a science *of* or a science *to* design. One form of the protest goes along the following lines:

Design, broadly speaking, is concerned with the invention of the forms of artifacts intended to meet certain goals or objectives. Suppose that a "client" (an individual, an organization, or an entire community) perceives a "need" or a "want" and has gone as far as establishing a specification of "requirements" for the desired artifact. A "designer" (again, an individual or a firm) is appointed to carry out the design.

Clearly, it is reasonable to claim that the design activity would only be initiated if no existing artifact *exactly* satisfies the given requirement. *To design is to invent.* The level of inventiveness may, of course, vary considerably. It will be quite low when the general structure or composition of the artifact is already well established or "mature" and the task of design is to fine tune the overall form of the artifact or make relatively small changes to meet the client's particular needs. On the other hand, the inventiveness may be of a very high order, as when the conceived artifact is sufficiently original in form, function, or some other characteristic as to be patentable or to give rise to an entirely new species of artifact.

Regardless of where in the continuum of inventiveness a designed artifact falls, to a greater or lesser degree, *every act of design is an act of creation*. It brings forth an entity which, in one way or another, is original. How then, it might be asked, can one explain, describe or prescribe such acts of creativity within a scientific framework?

Another line of argument draws upon the widely observed phenomenon that two or more independent designers posed with the same problem will very often produce quite different solutions. Indeed, this is why architectural competitions are held for the purpose of selecting a design for, say, a major new public building. There is, then, a strong element of *uniqueness* characterizing the output of design thinking, a uniqueness that is as much the fruit of the designer's particular style of thinking, judgement, and taste as of his knowledge and intellectual capacity.

How, one may again ask, can we ever construct a scientific understanding of such processes? Uniqueness is, after all, the hallmark of the arts and of artistic expression. Surely, the sceptic will argue, it makes far more sense to talk of the *art* of design than of its science, if only to pay explicit homage to the nonpredictive, contingent, subjective, idiosyncratic nature of this activity. This was the kind of thinking that prompted Donald Knuth, arguably one of the two or three most influential figures in the intellectual development of programming technology, to name his remarkable series of texts *The Art of Computer Programming*—notwithstanding the deeply mathematical and analytical nature of their contents.[5]

In these pages, I hope to respond to questions such as these. More precisely, the aim of this book is to show that despite its wonderfully contingent and nonpredictive nature, despite the creativity it entails and the uniqueness of its products, despite the ubiquitous intrusion of values pertaining to ethics, aesthetics, ideology, and society, *the processes by which artifacts are invented and designed—which we may legitimately regard as the intellectual epicenter of technology—exhibit certain significant features or attributes that appear to be universal and timeless in their nature.*

This, then, constitutes the central theme and thesis of this book. Among its implications are two that stand out.

One is that if indeed technology exhibits features that are universal and timeless—that is, they hold across the different artifactual domains and historical time, then it needs to be uncoupled from natural or "fundamental" science in the sense that our understanding of how artifacts are created—our understanding of the nature of technological creativity—should not *have* to rely on our knowledge of the laws of nature. This follows from the fact that science is relatively young—originating, in its modern form, in approximately the fourteenth century (according to the historian Herbert Butterfield)[6]—whereas technology, if we are to believe anthropologists, predates *Homo sapiens*. More pointedly, for a large part of the history of humankind, technological development has been independent of the growth of scientific knowledge.[7] Thus, if we wish to understand how the Gothic cathedrals of the twelfth and thirteenth centuries were designed, we can scarcely refer to the "engineering science" of strength of materials, which was first developed in the time of Galileo in the early seventeenth century.[8] Likewise, if we wish to understand how much of computer software is designed and built, we cannot draw upon the mathematical theory of computer programs since this theory has hardly had any impact on software technology.

On the other hand—and this is the other implication of the thesis stated above—if design or the invention of artifactual form as a process does indeed exhibit universal regularities, it is the very presence of these regularities that makes design and invention a legitimate field of scientific inquiry—for, the uncovering of hidden regularity or order is what science is all about.

My aim in this book then, is to give an account of some of the alleged "universal and timeless" characteristics of the processes of design and invention. Let us, somewhat grandiosely but for convenience, call this account *a theory of technological creativity*.

Intended as it is to be a "scientific" account, we would like our theory of technological creativity (more accurately, the *beginnings* of a theory) to be systematic and empirical. By systematic we mean that the theory should be built in an incremental and connected way. By empirical we mean that the theory must be testable. Furthermore, if we expect the reader to acquire some degree of belief in this account, we must provide ample evidence in support of the theory. Thus, our account of the nature of technological creativity will be interwoven with descriptions of certain episodes in the history of technology—these latter serving as actual tests of the theory. We shall draw upon a diversity of artifacts as the basis of the tests, including computers, high-temperature alloys, Victorian bridges, computerized typesetting, the early telephone, airplanes, and even abstract artifacts such as design methods and computer languages.

Ever since Thomas Kuhn published his seminal book *The Structure of Scientific Revolutions* in 1962, we have all become ultraconscious of the importance of *paradigms* in scientific inquiry.[9] A paradigm, in Kuhn's sense, is a complex, integrated network of theories, models, procedures and practices, exemplars, heuristics, and philosophical assumptions that establishes and defines the framework within which scientists in a given community identify and solve problems. For example, almost all contemporary evolutionary biologists and paleontologists work within the paradigm constructed, over the past century or so, around Darwin's theory of evolution. And physicists of the latter half of this century investigate the nature of elementary matter within the quantum-mechanical paradigm.

So also, our account of how humans invent artifactual form is constructed within a particular paradigm in psychology. This is often referred to as the *information processing* or *symbol processing* paradigm and it forms the base of the young but burgeoning discipline of cognitive science. One might say, then, that the account of technological creativity to be put forth in this book is a *cognitive* theory.

What do we really hope to achieve with this sort of inquiry? What is the point of trying to understand technological creativity?

First, it appeals to our intellectual curiosity about the nature of creativity. Like art and science, the practice of technology is an integral part of human existence. And as artists, scientists, and technologists go about *their* proper businesses, as they paint pictures, write novels, resolve puzzles of nature, or invent artifacts, some of us *want to know* how such creativity comes about. What was the nature of the processes by which Picasso arrived at his *Guernica*? How did Kepler derive his laws of planetary motion? How was it possible for Charles Babbage to conceive the Analytical Engine? Even lesser acts of creativity than these provoke our curiosity. How, for example, do scientific and other kinds of knowledge interact in the design of complex artifacts such as bridges, computers, and alloys? What *are* these other kinds of knowledge? What kind of reasoning is entailed? What are the differences (if any) between scientific and technological creativity?

For most scientists, artists, or engineers, it is the product itself—the solution to be obtained, the novel to be written, the nature of the artifact to be designed—that is the focus of interest. For those interested in creativity, it is the mind of the producer that commands interest. In the particular case of technology, we wish to know what kind of thought process can lead to the translation of sometimes very abstract human needs and wants into specific artifacts that must interact with and function in the physical world.

There is a second reason, and this is of a more practical nature. Technology has its vociferous critics as well as its passionate supporters. In an age in which ecological and industrial disasters are beamed into our living rooms literally as they unfold, we are acutely conscious of the more spectacular failures of technology. At a more mundane level, we are aware (and perhaps resigned to) the failures of what the psychologist Donald Norman called "everyday things"— telephones, faucets, doors, videocassette recorders and similar artifacts that challenge and frustrate our everyday existence by failing to do adequately what they are supposed to do.[10]

The point is that technology or engineering is, in a rather fundamental way, a *cognitive* activity. Thus, it seems to me that we will gain a much deeper understanding of why or how artifacts fail—whether spectacularly, as in the case of bridges, spacecraft or nuclear power plants, or in the more insidious manner of everyday things—and of the intrinsic limits of technology if we have some understanding of how artifacts are created by and in the human mind. We tend to couple the triumphs and failures of technology with those of the underlying theories. We forget that these successes and failures are as much reflective of the power and limitations of our cognitive capabilities. To understand technology is to understand how humans think, use knowledge, and reason. Or, as the Caribbean journalist and writer C. L. R. James might have said, What do they know of technology who only technology know?[11]

There is yet a third and still more practical reason for an inquiry such as this. It has to do with the fact that with the remarkable developments in computer science and its applications in recent years, computers are now increasingly being expected to carry out some of the "higher" mental functions such as design. Clearly, any understanding we achieve of the cognitive nature of technological creativity will both enhance the capabilities and define the limits of "computer-aided" or "knowledge-based" design systems.[12]

2

Artifacts: Material and Abstract

On the back cover of a reissue of George Sturt's classic *The Wheelwright's Shop*, the book is described as offering "a unique glimpse into the working lives of craftsmen in a world since banished by technology."[1] We see here a distinction being made between *craft* and *technology*—a distinction not uncommon since, according to current usage, craft involves the exercise of special skills, largely manual, by individuals, whereas technology implies the use of science, machines, and mass manufacture. Yet it is worth remembering that the word *technology* comes from the Greek τε΄χνη meaning "art" and "craft."

The separation of craft from technology is no doubt useful to social, economic, and cultural historians, but for the purposes of this book it is their commonality which is of interest. The craftsman of old, laboring in his shop, and the modern engineer sitting at her computer workstation seem worlds apart; and yet they share this one central thing: they are both concerned with the *creation of artifacts*. They are both, in this particular sense, practitioners of technology. Broadly stated then, *technology* will be used here to mean, collectively, the identification of practical needs and wants and the conceiving of artifacts in response thereto, the detailed development of artifactual form, and the implementation of such forms in the shape of actual artifacts.

In common parlance, *artifacts* are what one sees in museums of archaeology and anthropology as relics of historically distant civilizations—the "rude products of aboriginal workmanship as distinguished from natural remains."[2] More literally, though, an artifact is, quite simply, *an artificial product*—an entity which is not an element of the natural world but produced or consciously conceived in response to some practical need, want, or desire. Artifacts are, thus, *utilitarian* in scope: they are intended to serve some practical purpose.

There are, of course, many types of non-natural entities that are not usually regarded as being utilitarian—paintings, sculptures, and jewelry come to mind; thus, utility or practical purpose serves as a convenient criterion for demarcating artifacts from art. According to this criterion, then, it seems reasonable, for instance, to think of architecture as a branch of technology rather than of art. But what constitutes utility is itself a moot point: A painting may be hung on a living room wall as much to fill an empty space as to give aesthetic pleasure; or jewelry may be made less to adorn than as a convenient means of investing in gold or precious stones.

The line between artifact and nature may also be fuzzy: A branch from a tree becomes a walking stick, a large rock or boulder a shelter in a rainstorm; a field becomes grazing pasture for cattle. Utility, like beauty, lies often in the eye of the beholder.

Material Artifacts

The vast majority of artifacts that are built and that we come in contact with are *material* in nature. The diversity of such artifacts in type and complexity is truly astonishing—no less than the variety of species known to exist in the living world. Examining the patent literature in the United States alone, the historian George Basalla has concluded that technological diversity is of the same order as that in the biological realm.[3]

As I write these words, in my immediate vicinity I see the following artifacts: a desk, an "executive chair," a coffee table, two visitors chairs, a fountain pen, several ballpoints, pencils, a letter opener, writing pads, a telephone, a stapler, paper clips, books, periodicals, a briefcase, file folders, envelopes, postage stamps, a typewriter, bookshelves, filing cabinets, window blinds, a bulletin board, thumbtacks, a computer, a laser printer, an electric kettle, coffee mugs, a photocopier, and a calculator.

Some of these objects clearly are quite closely related to one another both in function and form—the fountain pen and the ballpoints, for instance, or the different kinds of chairs; others are also related but in more complex and not easily discernible ways as, for example, computer, calculator, and typewriter. Such artifactual family resemblance of varying degrees on the one hand, and sheer artifactual variety on the other, make it tempting to think in terms of "species" and "genera" of artifacts—of "evolution" in the technological domain. Whether one should treat the idea of artifactual evolution as simply a fanciful metaphor or whether it is useful to interpret it more literally will be discussed later in this book.

Inner and Outer Environments

In *The Sciences of the Artificial*, Herbert Simon made the point that material artifacts exist at the interface of what he called *outer* and *inner* environments.[4] By the former, he meant the physical environment in which every material artifact must reside and which obeys the physical laws of nature. The artifact, internally, is composed of matter which is also subject to natural laws. Such matter with its physicochemical properties constitutes the artifact's inner environment.

Here, then, lies the main intellectual problem of technology. Both the inner and outer environments of material artifacts obey nature and exhibit properties dictated by nature. The raw materials of the craftsman and the modern engineer alike are, in fact, these natural environments. The task of technology is to invent artifactual forms and make artifacts in the images of such forms such that the inner and outer environments of each artifact cohere and cooperate in order to satisfy some given set of human wants. The task of technology is as much to overcome nature's perversities as to exploit its geniality. The history of the development of an artifact as it is designed or as it changes over time is to a large extent the history of how different aspects of the natural world—the artifact's two environments—are teased into a state of mutual cooperation.

Consider the humble lead pencil, the history of which has been recently written with such grace and erudition by Henry Petroski.[5] Its inner environment consists of graphite, clay, cedar wood, and other materials. Each has its own idiosyncrasies: graphite is brittle, porous, hard; clay is plastic, mixes well with water and, when combined with graphite, affects the hardness of the latter; cedar wood, in long, thin, columnar pieces will snap when force is applied at its ends. Graphite, clay, wood, and all the other substances that go into the making of the pencil have intrinsic properties that make them respond in characteristic ways to the forces or pressures imposed upon them. They are matter.

The pencil itself is not matter, at least in the above sense. Artifact is nature disguised. We do not consciously see the pencil as graphite, wood, and clay. We see it as a writing instrument, the form of which allows it to be held by one's fingers and which makes dry, dark, erasable marks on paper by exerting a moderate amount of pressure. These are its *artifactual properties*—the properties that make the combination of graphite, clay, and wood into a pencil. The form or shape of the pencil mediates, on the one hand, between graphite, wood, clay, and their respective properties—its inner environment as Simon would say—and, on the other, the properties of plane surfaces (color, texture, resistance to imposed pressure) and the anatomy of human fingers; the latter elements, collectively, constitute the pencil's outer environment.

Abstract Artifacts

If it is accepted that any utilitarian man-made entity is an artifact, one must also accept that artifacts do not always *have* to be materially or physically tangible in form. The military commander constructs a *strategy* for the conduct of a campaign; a committee seeking to bring a future Olympic Games to its city proposes a detailed *plan* for the location, logistics, and resourcing of the Games; a software engineer proposes a new *method* for managing the life cycle of software systems; a management consultant develops a new *organization* for a troubled corporation; a board of trustees approves the specifications of the *architecture* of an art museum's new wing; a computer scientist invents a new *algorithm* for multiplying two matrices.

Strategies, plans, organizations, architectures, designs, algorithms, and methods are all examples of *abstract artifacts*. They are abstract in that they are not material entities—they are literally intangible. One can touch a building or a computer. One can touch neither the architecture of that building nor the organization of the computer. Yet such entities are indubitably artifactual—they are useful, man-made constructs intended to satisfy certain human goals. Furthermore, they are objectively real, for once created, they can be used, communicated, analyzed, and modified by anyone who cares to do so.

The philosopher Karl Popper has suggested that in addition to the traditional Cartesian categorization of the universe into objective physical reality and subjective mental events—which he termed "World 1" and "World 2" respectively—there is a third class of entities that are neither physical nor mental; they are, rather, the abstract but objective *products* of mental processes—entities such as theories, laws, and ideas. These belong to what Popper called "World 3."[6] Thus, while material artifacts clearly belong to Popper's World 1, abstract artifacts are elements of World 3.

Perhaps the most obvious and important characteristic of abstract artifacts is that they are rendered visible through symbols or the combination of symbols—that is, by *symbol structures*. The design of a bridge or machine is expressed by means of engineering drawings augmented by text and, possibly, mathematical equations; algorithms are described using a mix of natural language, mathematical symbols, the notation of computer languages, and flowchart diagrams; the architecture of a building is explicated through architectural and perspective drawings; a military campaign is made visible using maps, diagrams, and text. Symbol structures are what make abstract artifacts objective.

A Distinction between Craft and Engineering

The creation of most material artifacts of any reasonable order of complexity—a bridge, an aircraft, or a computer, for instance—necessitates a design, or "conceptualization" phase that precedes the manufacture, or "making" phase. A design, expressed in the form of an engineering drawing, for example, is itself an abstract artifact. Thus, most acts of technology entail the production of abstract and material artifacts, the former being the artifactual form, the latter the artifact itself, and the first preceding, determining, and constraining the other.

History tells us that this separation of conceptualization and making has not always been the case—indeed, one way in which we can distinguish between craft and technology or engineering (as the latter terms are commonly understood) is, as Christopher Jones noted, the fact that the craftsmen of old bypassed the conceptualization or design stage.[7] More precisely, for the craftsman, there was no *independent* conceptualization stage the output of which was a "design." Conceptualization and making were inextricably intertwined. The craftsmen did not externalize their ideas in the form of symbolic descriptions; the externalization lay in the material artifact itself.

A very similar idea is expressed by Christopher Alexander in his *Notes on the Synthesis of Form* where, in the specific context of buildings, he writes of the *unselfconscious* process by which buildings are brought into existence in highly traditional cultures; in such a process, Alexander notes, there is little explicit thought about what rules to apply—the notion of what to do is embodied in the doing itself. Furthermore, there are no external means of communicating ideas.[8]

It is not particularly easy to summon documented evidence in support of this suggested attribute of craft, for it is a notorious and unfortunate fact that the technologist, traditionally, has been loathe to reflect in writing on the nature of his work. One can hardly fail to notice, for example, that in Brewster Ghiselin's *The Creative Process*, an anthology of writings by highly creative people about their own acts of creation, not a single inventor or engineer appears.[9] Certainly, few engineers have been so richly explicit or generous in this regard as were the Victorians William Fairbairn and Edwin Clark on the design and construction of the Britannia Bridge in North Wales—a generosity I have gratefully exploited later in this book when the design of this famous bridge is considered in some detail.[10] One plausible cause of the poverty of written records of the technological process is that drawing rather than writing has been the preferred medium of thought and expression for an engineer.[11]

If engineers are thought to be taciturn, then still more so were the craftsmen. No doubt this was due partly to the fact that craftsmen were in the main unlettered and uneducated. Another factor is the secrecy and mystique that shrouded each trade or craft, a mystique nurtured assiduously by the trade guilds. A remarkable exception to this tendency was George Sturt, whose *Wheelwright's Shop* was briefly mentioned at the start of this chapter. Sturt's book was published in 1923, but it describes, in autobiographical form, the nature of the wheelwright's craft for the period 1884–91, during which he ran the business inherited from his father.

If *The Wheelwright's Shop* is unusual in what it reveals about the nature of a particular craft, this is no doubt owing to the fact that Sturt was no rude, unlettered mechanic or businessman of his time. He was a writer who contributed to a socialist journal, was influenced by Ruskin, Thoreau, Emerson, and Walt Whitman, and corresponded with the novelist Arnold Bennett.[12]

Sturt described every phase in the making of wheels for the different kinds of vehicles in demand within the rural and country town community that provided his custom and of which he was a part—from the buying of timber to the tiring of the wheel. The significant word here is *making*—the creation of the material artifact itself. No mention is made of design or even the preparation of simple drawings. The closest the wheelwright gets to a conceptualization stage, to the construction of a preparatory abstract artifact, is when timber already bought is to be selected for sawing. Sturt describes how the wheelwright would pencil in possible shapes of a component of the wheel, called the felloe-block, on pieces of timber and "play" with alternative shapes to match the curve of the wood.[13] This was, however, a rare activity in the whole enterprise. Design as an independent phase was hardly necessary.

> What we had to do was to live up to the local wisdom of our kind; to follow the customs, and work to the measurements which had been tested and corrected long before our time in every village shop all across the country. . . . A good wheelwright knew by art but not by reasoning the proportions to keep between spokes and felloes; and so too a good smith knew how tight a two-

and-a-half inch tyre should be made for a five foot wheel and how tight for a four-foot and so on. He felt it in his bones. It was a perception with him. . . . Every detail . . . had to be learnt either by trial and error or by tradition.[14]

But the craft tradition, even when alive and well, was but one face of the technological process. Design and the separation of conceptualization from making become *psychologically* necessary when the size and complexity of the artifact exceed the cognitive capacity of an individual mind. Conceptualization is a conscious cognitive process and psychologists tell us that for such acts to be performed, the requisite knowledge must be contained in the person's short-term, or "working" memory. And as George Miller famously revealed in the 1950s, the number of meaningful items of information—"chunks"—that can be simultaneously held in working memory is disconcertingly small.[15]

Thus, it is not that the craft way of making artifacts has been supplanted by conceptualization-followed-by-making. Observing, for example, skilled programmers at work at their computer terminals, we cannot fail to be so often struck by the uncanny unselfconsciousness, in Alexander's sense, of this most modern of technological processes. Rather, the craft and design approaches have coexisted across the centuries. Perhaps the earliest identifiable theorist of building design is Vitruvius. His *De Architectura*, which appeared around 25 B.C., presented what we now unequivocally would call design knowledge and design procedures.[16] And yet Vitruvius was codifying knowledge and a theory of building and architecture that had developed as part of what William Addis referred to as the "Greek design revolution" over several centuries—from about the seventh century B.C.[17] Such knowledge of procedures, algorithms, and rules of proportion enabled the master builders of the Hellenic (c. 800–323 B.C.) and Hellenistic (323–31 B.C.) periods to plan their buildings, visualize how they would appear well in advance of when they would be built, justify their designs to themselves and to their clients, and communicate the design to others, especially the workmen. For the Greeks, the separation of conceptualization from making was a fact of life. And, by the time of *De Architectura*, the architectural drawing—an abstract artifact par excellence—was very much a part of the architectural Zeitgeist.[18]

The Manufacturing Process as Artifact

In the realm of such technological disciplines as mechanical engineering, metallurgy, and chemical engineering, one of the most ubiquitous of entities is the *manufacturing process*. A process is a specification of how to proceed in order to achieve some practical end. In this sense they resemble algorithms; and like algorithms, they are manifested in some textual or pictorial language, that is, they are abstract; furthermore, processes have objective reality in that they can be communicated between people, examined, criticized, and modified. In all these respects, processes are abstract artifacts.

Yet in one way they resemble material artifacts: most manufacturing processes when newly invented can be patented. In the history of technology, processes, in spite of their abstract, symbolic form, have been treated by their creators and society alike as though they were material in nature.

The key to this apparent paradox appears to lie not in the abstract nature of the manufacturing process-as-artifact itself, but in the character of the environments of processes. To understand this, consider the following example.

One of the first methods for the manufacture of wrought iron using coal rather than wood charcoal, due to the brothers Thomas and George Cranege, was an early form of what came to be called the "puddling process." The process was developed in the famous Coalbrookdale iron works founded in Shropshire, England, by Abraham Darby in 1709, and was granted a patent in June 1766.[19] The Cranege process was described in a letter dated April 25, 1766, by Richard Reynolds, then head of the Coalbrookdale works, to an associate in which, after declaiming that the process was "one of the most important discoveries ever made," he requested his friend to take out a patent for it without further delay. Reynolds then went on to say, "The specification of the invention will be comprised in a few words, as it will only set forth that a reverberatory furnace being built of a proper construction, the pig or cast iron is put into it, and without the addition of anything else than common raw pit coal, is converted into good malleable iron, and being taken red hot from the reverberatory furnace to the forge hammer, is drawn out into bars of various shapes and sizes, according to the will of the worker."[20] According to the Victorian biographer Samuel Smiles, this same wording was used in the patent application.[21]

If we pause to consider the specification of this process as just quoted, we see that it consists of a set of symbols designating a variety of operations: *placement* of pig or cast iron on the hearth of a reverberatory furnace, *addition* of raw coal, *operation* of the reverberatory furnace,[22] *removal* of the purified iron thus obtained, and *drawing out* of the bars. Thus, the inner environment of the Cranege process consists of physical, mechanical, and chemical operations all subject to natural laws and observable physicochemical behavior. The outer environment, of course, consists of pig or cast iron, coal (and its various chemical constituents), the furnace—a material artifact with its own characteristic properties—the flame and gases which heat the material, and the wrought iron itself. It is little wonder that processes such as these are viewed in the same way as material artifacts insofar as patent laws are concerned. The abstract quality of the process-as-artifact is masked by the material nature of its environments.

The Janus-Faced Nature of Computer Software

In June 1949, the EDSAC, designed and built by Maurice Wilkes and his colleagues in Cambridge, England, became the world's first fully operational stored program computer.[23] This event marked the birth of a type of artifact never before encountered in the long annals of technology. We now call this artifactual class *software*. The individual artifacts belonging to this class are, of course, computer programs.

The uniqueness of software artifacts is an empirical fact—prior to its invention there had never been anything like it in technological history. To be sure, there had been ingenious automata of various sorts, but insofar as such devices were "programmed," the programs were an integral part of—were "hardwired" into—the devices. Perhaps the one genuine precursor to software lay a century before in the work of the astonishing Charles Babbage (1791–1871) who in conceiving the Analytical Engine seems to have anticipated the *idea* of what later came to be called the computer program. The Analytical Engine, however, was doomed to failure for a variety of reasons.[24]

A large measure of the distinctiveness of software lies in its Janus-like nature: On the one hand, a vast corpus of precise knowledge about software has grown around the notion of software as a wholly abstract, indeed mathematical, entity. Computer programs, according to this view, are mathematical structures, like the theorems of algebra and geometry, and just as one can formally *prove* a theorem so also one can *prove* that a program is correct. Software, in this sense, is a purely abstract kind of artifact.[25]

There is, however, a complementary view which argues that software attains its artifactual nature—that is, becomes a utilitarian object—only when it resides in a computer's physical memory and is actually executed by the computer's electronic circuits. A computer program displayed on the terminal screen of a personal computer or listed on a printout is a piece of abstract text. It is simply a formal procedure or algorithm, though expressed in a very precise, unambiguous language. The program becomes a useful *computational artifact*—that is, software—when it is in an executable form for a particular computer or type of computer; the artifact is the program itself *in conjunction with* the physical resources (for instance, memory and the circuitry) and any other resources that enable the program to be actually used.

A software system or device is, then, a *causal* entity in that it enables physical events to occur: signals to be transmitted along wires, states of semiconductor memory to be altered, marks to be printed on paper, switches to be turned on or off, and so on. Furthermore, the actions performed by software consume *physical time*; and the software itself occupies *physical space* (in memory). The operations out of which a particular program is built are not purely symbol-processing operations; they cause changes in physical states over physical time and may even, physically, malfunction. Thus, the inner and outer environments of software are of the same nature as those of a manufacturing process: they are subject to physical laws.[26]

This Janus-like character of software is not merely a matter of academic curiosity. That software has a material base has come to be quite explicitly recognized by the laws which now allow software to be patented. It has also had a substantial influence on the education and training of software practitioners: On the one hand, adherents of the software-as-abstract-artifacts school of thought teach and expound on programming as if it is a branch of mathematics;[27] on the other hand, courses and texts on what has come to be known as "software engineering" treat the topic in terms that make it virtually indistinguishable from the disciplines concerned with traditional material artifacts.[28]

And, unfortunately, like East and West in Rudyard Kipling's poem, the twain refuse to meet.

The Fallibility of Artifacts

The notion of the artifact as an interface between two environments, each possessed of its own characteristic set of properties, provides an important key to understanding something about the intellectual problem of technology in general and of the invention or design process in particular. It suggests to us, for instance, why technology in its modern sense is so widely regarded as the *application* of science.[29] When we talk of a particular technology having a scientific basis, what we are implying is that we have acquired a relatively deep understanding of the inner and outer environments pertaining to the artifacts of interest. And, the deeper this understanding of the environments—the more we know and understand the science underlying the artifact—the more "progress" will we have made in that technology. Or so the wisdom of our times tells us.

This belief is to some measure justified—as long as one never forgets that the "scientific" understanding of inner and outer environments has, historically, been a condition *neither necessary nor sufficient* for the creation of satisfactory artifacts: Gothic cathedrals were being built long before the theory of structures had emerged, while the development of a mathematical theory of programming has had, regrettably, little impact on the engineering and quality of software.

What is, perhaps, far less appreciated is that the concept of the artifact as an interface between two environments also gives us insight as to *why artifacts are so fallible*—even when the science behind the artifact is well understood. "Nothing we design ever really works," as the industrial designer David Pye put it.[30]

To see why this is so, we must realize that when an artifact is being created, the technologist's awareness and comprehension of its inner and outer environments are inevitably *selective*—for a variety of reasons, part cognitive and part having to do with the state of knowledge about the relevant environments. Science itself is selective; the scientist tackles and attempts to solve only those problems that, in the phrase of the immunologist and philosopher Peter Medawar, are soluble.[31] Unfortunately, an artifact, even one conceived in, and informed by, the most exact of sciences—that is, one whose inner and outer environments are well understood—is no respecter of either the limitations of knowledge or the limits to our cognitive abilities. An artifact, once created, acquires a life of its own, not in any obscurantist sense but simply because, like natural organisms, it is governed by the principles and properties of its own inner and outer environments; and these may well be far beyond the creator's anticipation, understanding, knowledge or cognitive capabilities.

The Britannia Bridge, built by Robert Stephenson and his associates in the 1840s was hailed as a revolutionary landmark of British engineering in the halcyon days of the Empire—a "national monument" according to one eminent Victorian.[32] It will loom large in this book, for its history and circumstances yield rich pickings on how an artifact of some considerable complexity can come into being. But we owe it to the civil engineer Henry Petroski for nudging our attention to the fact that this bridge, designed to carry trains across the Menai Strait in North Wales, was as much a failure as it was a spectacular success.[33] The failure lay in the engineers' limited understanding of the bridge's actual outer and inner environments.

The Britannia Bridge (Figure 2-1) was, in effect, a huge tubular wrought-iron beam through which trains would pass. For its inventors—and it was an invention of the first order not merely a design, as we shall later see—the relevant inner and outer environments pertained to the universe of load-bearing structures; the relevant laws were those governing how beams of various shapes behave under load. And, as the documentary records of the project show, when the state of analytical structural knowledge of the period failed to yield the requisite insight, Stephenson and his associates, in particular William Fairbairn, in exemplary scientific fashion, conducted extensive and detailed experiments that generated new knowledge about the bridge's *anticipated* environments.[34]

In terms of the relevant physical inner and outer environments as the engineers then understood it, that is, solely in *structural* terms, the Britannia Bridge achieved the goals established for it very satisfactorily; indeed, it remained operational for some 130 years until, in 1970, an accidental fire partially deformed it. But the actual inner and outer environments of this bridge were constituted of more than mechanical loads and wind forces. They included the excessive heating of the wrought iron by the sun; the smoke and soot expelled by the steam locomotive but entrapped within the closed tubular form that was the bridge; and the human passengers and the acute discomfort they would be subject to by heat, smoke, and soot. Robert Stephenson and his associates failed to anticipate that these factors were as much part of the bridge's outer environment as was the load imposed by the weights of the bridge itself and the train, and the forces exerted by wind.

This failure—literally an environmental failure in the customary modern usage of this term—was caused by what the psychologist James Reason in his recent study of human error would classify as a *knowledge-based mistake.*[35] This is one of several types of errors Reason and others have studied, and is generated in the realm of conscious, knowledge-intensive problem-solving activities when there is either a failure or deficiency in the selection of a problem-solving objective or when the intended problem-solving action or plan fails to achieve the intended goal—provided that these failures are not due to the intrusion of some chance factor.[36]

Artifacts may also fail because of other forms of error. However, insofar as technology is concerned with the purposeful, conscious creation of artifacts intended to meet desired human goals, knowledge-based mistakes are by far the most significant causes of why artifacts are so fallible. And, as previously noted, such errors and the failures that they spawn are due in part to cognitive limitations and in part to the inadequacy of available knowledge concerning the actual (rather than postulated) environments of artifacts once they come into being.

Fig. 2-1. *The Britannia Bridge.* It was designed and built by Robert Stephenson and his associates between 1844 and 1847, and spanned the Menai Strait in North Wales at the location known as the Britannia Rock. In essence, the bridge was a wrought-iron rectangular tube through which trains could pass.

3

The Birth of Technological Problems

In the Musée Picasso in Paris, there is a sculpture which, from a distance, is obviously a bull's head. On closer view, it resolves into a bicycle seat attached to a handlebar. The spectator can scarcely fail to marvel at the impudent simplicity of this work and to reflect, even if only briefly, on Picasso's capacity to see a bull's head in an object as mundane as a bicycle.

Yet, it might be argued that Picasso perhaps saw a bull's head in a bicycle because he was *searching* for this form in something that was patently otherwise. That is, he was seeking a solution to a particular problem, namely, the representation of the head of a bull in an unusual medium. The desired shape or form was already in his mind, and when he chanced to see a handlebar and a bicycle seat, an analogy was formed between these objects and the shape occupying his thoughts. Fortune, as Louis Pasteur put it, favors the prepared mind. Picasso's "problem" itself helped determine the solution.

Whether or not this is a plausible explanation as to the origins of this particular sculpture, or, more generally, whether works of art can be explained as solutions to specific problems—as was suggested by the art historian E. H. Gombrich in his *Art and Illusion*[1]—is a matter we shall leave to the student of artistic creativity to determine. But there seems little doubt that technology is the quintessential problem-oriented activity. Artifacts are responses to specific problems. To state this is, perhaps, to state the obvious and yet it helps to focus attention on the very beginning of the act of technological creation, namely, the birth of technological problems. How do such problems originate?

Need

Necessity, said an eighteenth-century playwright, is the mother of invention. Whether necessity is the mother of *all* invention is certainly questionable; there is little doubt, though, that many technological problems are sired by need or necessity—that is, "circumstances requiring some course of action."[2] Need and want are not entirely synonymous; a want may indeed stem from necessity but it can also arise from desire or a sense of dissatisfaction with the way things are. As we shall see later, dissatisfaction is the second major source of technological problems.

The nature of need varies, of course. In the 1930s, the development of the gas turbine gave rise to the need for an alloy to make turbine blades. The alloy would need to resist creep deformation and creep rupture at high temperature,[3] corrosion due to hot gases, and accidental shocks, and yet would have the capability of being mechanically formed.[4] Necessity was, thus, certainly the mother of invention of the first "superalloy." It was, furthermore, a purely *technological* need, being the outcome of a separate technological problem—the development of another artifact, the gas turbine. No superalloys, no gas turbine.

In a similar way, the problem of developing a stored-program digital computer in the 1940s created, in turn, the need for a device capable of storing "large" amounts of information—that is, a large-capacity memory device.[5] Such a memory was necessary if the stored-program computer was to become a reality: No large-capacity memory, no stored-program computer. The ultrasonic delay line memory invented by Presper Eckert and John Mauchly in Philadelphia, and the electrostatic storage tube developed independently by F. C. Williams in Manchester and Jan Rajchman in Princeton were the immediate responses to this particular problem. As in the case of the first superalloy, the problem of the electronic storage device was spawned by a purely technological need—one artifactual problem giving rise to another.[6]

The first successful use of coke to smelt iron is a rather significant episode in metallurgical history. According to Samuel Smiles's account, the process was invented by Dud Dudley in Worcestershire, England, in 1620.[7] Prior to this time, the common fuel used for iron making was wood charcoal. However, severe restrictions, imposed by legislation, on the use of wood in iron making directed the attention of the iron makers of the time to the employment of "pit coal" as a substitute.[8]

Dudley's process was, thus, a response to a need to use an alternative fuel to wood, specifically coal, for iron making. The nature of this need, however, differs from those of the gas-turbine alloy and computer memory: there, the development of one relatively new artifact gave rise to the necessity of inventing another; in the case of iron making, the artifact itself, namely manufactured iron, was not new. Rather, it was ensuring the continuity of its production and availability which necessitated the invention of a new artifact, namely, a process for iron making using coal.

The origin of the Britannia Bridge, mentioned briefly in the preceding chapter, represents yet another variety of need. In July 1844, the British Parliament passed a bill authorizing the Chester and Holyhead Railway to build a railway route across the Menai Strait in North Wales. This was to be one of several critical links in the railway route connecting London to the mail boats crossing over to Dublin.[9] The problem that eventually led to the design and construction of the Britannia Bridge was, thus, based undoubtedly on a need. It was a circumstance demanding a course of action. Without the railway lines crossing the Menai Strait (and, incidentally, other waterways), the effectiveness of the railway network across Britain would be diminished.

However, the circumstance here differs from the earlier ones in that it is a need born not from technology itself, as was the case with the turbine blade superalloy and the ultrasonic computer memory, nor from ecological concerns, as was the circumstance that led to Dud Dudley's process; rather, its origins lay in the demands imposed by commerce, communication, and transportation.

Dissatisfaction

Among the many new artifacts that resulted as byproducts of the Second World War, is the general-purpose electronic digital computer. The first such computer, the ENIAC, was conceived and built between April 1943 and February 1946 by a group led by Presper Eckert and John Mauchly at the Moore School of Electrical Engineering of the University of Pennsylvania.[10] The ENIAC, like its contemporaries radar and the atomic bomb, was a technological response to specific problems precipitated by the war. Its primary purpose was the calculation of "firing tables"—that is, tables of ballistic trajectories.[11] The main sponsor of this project, the Ballistic Research Laboratory, however, hoped for other uses also; for this reason the ENIAC was designed as a general-purpose calculating device rather than a specialized artifact which would only compute firing tables.[12]

We may think that the ENIAC was born of a need; but it would be more accurate to say that it was an outcome of another source of technological problems—perhaps the most fertile of psychological sources—namely, *dissatisfaction* with the way things are and the consequent *desire* or *want* to correct or improve the state of affairs.

For, it was not that ballistic firing tables were not calculable prior to the ENIAC. An electromechanical device called the differential analyzer was already in use. However, as Arthur Burks, a key participant in the development of the ENIAC and its successors has recorded:

> One day John [Mauchly] suggested to Herman [Goldstine, the army lieutenant in charge of the calculation of the firing tables] that trajectories could be calculated much faster with vacuum tubes. Herman thought John's suggestion a good one, and asked for a proposal. John, Pres. Eckert and J. G. Brainerd then wrote a report on an Electronic Diff* [sic] Analyzer (2 April 1943). This was submitted to the Ballistics Research Laboratory. . . .
>   In this proposal the Moore School offered a machine that would compute ballistic trajectories at least 10 times as fast as the differential analyzer and at least 100 times as fast as a human computer with a desk calculator. Herman persuaded the Ballistics Research Laboratory to fund the proposal. The electronic diff* analyzer later became the Electronic Numerical Integrator and Computer, or ENIAC for short.[13]

The immediate factor that gave rise to the development of the ENIAC was dissatisfaction with the prevalent state of affairs in the calculation of firing tables, and a desire to improve this state of affairs. This same sense of dissatisfaction, *but this time with the* ENIAC *itself* marked the onset of the next major landmark in computer technology, namely, the invention of the logical principles and architecture of the stored-program computer between 1944 and 1946—an invention, incidentally, of an *abstract* artifact, the material realizations of which were to follow some three years later, in the universities of Cambridge and Manchester in England.

The historian of technology William Aspray has recorded that discussions about a new kind of computer had actually been initiated within the Moore School group even before the ENIAC had reached completion, and had been precipitated by several deficiencies recognized by Eckert, Mauchly, et al. in the ENIAC design.[14] The major problem with the ENIAC was the inefficient method for setting up a program prior to its execution, for there was no provision for what was then termed "automatic programming" of the ENIAC.[15] One had to set up a program manually, by plugging in cables and setting switches.[16] The problem of designing what came to be known as the stored-program computer, thus, arose because of a basic dissatisfaction on the part of the ENIAC group with the operating characteristics and design of the ENIAC. As mentioned in the previous section, this desire to develop a new kind of computer led, in turn, to a need to invent memory devices with size sufficient to store programs. Dissatisfaction at one level produces need at another.

The evolution of the computer reveals many other examples of the role of dissatisfaction in the birth of technological problems; it is, perhaps, worth pursuing this matter a bit further by examining one other major development in the very early days of the digital computer. This is the invention, in 1951, of microprogramming.

Maurice Wilkes (b. 1913) was one of the young pioneers in the practical development of the stored-program digital computer. The EDSAC was designed and constructed by Wilkes and his collaborators between 1946 and 1949 at what was then called the University Mathematical Laboratory in Cambridge, England. The EDSAC was, in fact, the first *fully operational* realization of the principles of the stored-program computer conceived earlier by John von Neumann, Presper Eckert, John Mauchly, and their colleagues in the United States, though only just ahead, in this respect, of the Manchester MARK I.[17]

Soon after the EDSAC was completed, Wilkes became concerned with what he perceived was an unsatisfactory feature of the EDSAC: the irregular, ad hoc, and, consequently, unduly complex structure of the EDSAC's *control unit*, the component responsible for controlling and sequencing the execution of a program by the computer. These characteristics were unsatisfactory since they made the design, repair, and maintenance of the control unit difficult.

Thus was born, again out of dissatisfaction with the existing state of affairs, a technological problem, which eventually led to Wilkes's invention of the principle of microprogramming and, later, to the design and implementation of practical microprogrammed control units.[18]

This problem, incidentally, was basically of an abstract or conceptual kind— it pertained to such abstract properties as "regularity" and "complexity."[19] Indeed, Wilkes has himself remarked that without a particular philosophical perspective on how the organization or structure of a computer ought to be, the problem he identified would not make sense.[20] Wilkes as much invented a problem as the solution to that problem.

One of the deities in the pantheon of American technology is Alexander Graham Bell (1847–1922), professor of elocution at Boston University, teacher of the deaf, and, of course, inventor of the telephone. Bernard Carlson, a historian of technology, and Michael Gorman, a psychologist, both of the University of Virginia, have recently constructed a fascinating and illuminating cognitive account of Bell's development of the telephone between 1874 and 1877.[21] For our present purposes, the relevant aspect of their account is the light it sheds on how dissatisfaction and the desire to improve influenced the genesis of what came to be known as the telephone.

By the mid-1870s, the telegraph had been established as a major means of rapid communication. In 1872 Western Union, who virtually monopolized the technology in the United States, had adopted Joseph Stearns's "duplex" system—a scheme whereby two messages could be sent simultaneously through a wire. However, as the volume of messages being sent increased, Western Union sought schemes whereby multiple messages could be simultaneously transmitted. As Carlson and Gormon put it, "it was soon clear that fame and fortune awaited the inventor of a four- or eight-message system."[22]

Bell apparently read a newspaper account of Stearns's invention of the duplex scheme and was convinced that, using his knowledge of speech, acoustics, and an electromechanical apparatus for analyzing and reproducing vowel sounds constructed by the German physicist Herman von Helmholtz, he would be able to produce a multiple-message "speaking telegraph."[23] Thus, Bell's problem was the outcome of the telegraph industry's dissatisfaction with the state of the art, conjoined with Bell's personal desire to improve upon the prevalent technology. In Bell's case, this desire was augmented and strengthened by his personal specialized knowledge, which suggested that a solution could be found.

If we step back still further, a hundred years before the time of Bell, we come to the period when the Industrial Revolution was in full swing in Britain. This was the time when the Yorkshire city of Sheffield was beginning to establish itself as one of the world's epicenters of the iron and steel industry. One of its denizens was Benjamin Huntsman (1704–1776), who in the mid-1740s invented the method of making cast or ingot steel that came to be known as the "crucible process."[24] But before becoming a pioneer steelmaker in Sheffield, Huntsman was a clockmaker and skilled mechanic in the neighboring Yorkshire town of Doncaster, who made his own tools and the instruments needed for his vocation. However, he found the steel he used for this purpose, commonly called "German steel," far from satisfactory.[25] His dissatisfaction with the quality of his material must have been very keen, for it impelled Huntsman to begin and pursue the experiments that led, eventually, to the crucible process. Huntsman has thus been credited with laying the foundations of all subsequent steel ingot making processes and with advancing the place of Sheffield, Yorkshire, and Britain in iron and steel making.[26]

As a final example, William Addis, a structural engineer, has noted how in the domain of load-bearing structures, some of the most important developments have come about "as a result of a sense of dissatisfaction, especially with the power of a design procedure . . . to justify a proposed design or with the ability of a design procedure to yield a design which is economical or safe ... such feelings have their origin in the designer's intuitive understanding of structural behavior, action, and adequacy."[27] An instance from Addis's own domain is the circumstance that led to the invention and subsequent development of the method of plastic design (also referred to as limit-state design) of steel structures between 1935 and 1950 by the British structural engineer John Baker and others.[28]

The story, as told by Addis, is absorbing for it recounts how a paradigm in the sense of Thomas Kuhn's celebrated "scientific paradigm"[29] was established in structural design. However, our current concern is only with the birth of the problem that led to the development of this paradigm and to what Addis called the "plastic design revolution."[30]

Up to the early 1930s, steel-frame buildings were designed and constructed in England according to regulations drawn up (in 1909) by the London City Council. However, it became apparent to many engineers that these regulations were unduly restrictive and did not allow advantage to be taken of the more recent developments and improvements in structural steel. An organization called the Steel Structures Research Committee was set up in 1929, its brief being twofold: to review the existing methods of steel-structures design, and to recommend how modern theories could be applied to develop more efficient and economical design methods.

The brief itself, especially its second part, defined a technological problem. But what gave rise to it were a number of concerns felt by engineers, many of which pertained to the *empirical and irrational character* of the design procedures of the time. Many years later, John Baker in his now classic treatise *The Steel Skeleton* would write: "[The] review of existing regulations left a strong impression that the methods of steel framed building in common use had no firm rational basis, and that the wide range of constants representing loads and stresses made it difficult to justify even as empirical. It was clear that no advance could be made until the real behavior of this form of structure under load was understood."[31]

Here we have, then, the creation of a structural-engineering problem that sprang from dissatisfaction with the perceived irrationality of the existing state of affairs.

Curiosity

Curiosity may well have killed the cat but it has historically given life to the natural sciences. For science, at least in its more pristine form, when not tainted by economic or military imperatives or what the funding agencies will support, is concerned with *understanding* natural phenomena and with the *explanation* of puzzles, paradoxes, and anomalies. The scientist asks, Why is such and such the case? How can we explain such and such phenomenon? Curiosity is the fount of scientific problems.

In technology, the questions corresponding to those above would be, Can an artifact be built that will do such and such? or, more simply, Can an artifact do such and such? Can a computer think? Can we build a flying machine? Can a tower be built that is taller than any other existing freestanding structure in the world? One might call the artifacts resulting from questions prompted by curiosity exercises in invention for its own sake.

However, while curiosity is not usually prompted by need, it is difficult to entirely separate technological curiosity from dissatisfaction, for in asking whether an artifact can do such and such, one is implicitly recognizing that there is no existing means to do such and such—that there is a void. Thus, the technologist is voicing a want or desire to fill this void. Furthermore, there is some evidence that the curiosity voiced by such questions as whether an artifact can do such and such is not of the idle sort. Rather, the technologist-inventor is tacitly asking, "Can *I* build an artifact to do such and such?"

One might say, then, curiosity as a source of technological problems is an expression of a very *personal* sense of dissatisfaction, on the part of the individual, with the way things are in some particular artifactual domain, coupled with a sense of being *personally* confronted with a challenge that the inventor must meet. Thus, it seems reasonable to believe that Maurice Wilkes, being personally dissatisfied with the nature of existing control units and, consequently, pursuing what he himself admitted to this writer was a "private" problem, was very much driven by curiosity in the above-mentioned sense.[32]

For Jacob Rabinow, holder of well over three hundred patents, whose list of inventions includes such varied devices as safety mechanisms for rockets and bombs, a gyro control for guided missiles, and the first magnetic-disk file for computers,[33] problems become alive when "you want to do something better than it has been done before or that has never been done before."[34]

Similarly, Masaru Ibuka, co-founder of Sony Corporation and inventor of a multitude of artifacts in the consumer electronics domain (including the Walkman personal stereo), has talked of his passion for inventing arising out of a need to satisfy his curiosity: "I invent because making new things provides one of the biggest joys of my life—as satisfaction of my curiosity."[35]

The nature of this curiosity, namely a comingling of private dissatisfaction and private or personal challenge, is further exemplified by Donald Knuth's development of TEX, the computerized typesetting system that is now used widely in the typesetting of scientific, especially mathematical, literature.[36] Knuth has written: "The genesis of TEX probably took place on February 1, 1977 when I first chanced to see the output of a high resolution typesetting machine . . . this fine typography was produced entirely by digital methods; . . . I realized that a central aspect of printing had been reduced to bit manipulation. As a computer scientist, I couldn't resist the challenge of improving print quality by manipulating those bits better."[37]

More recently, in a communication to this author, he provided further insight on this matter:

> Later when I learned . . . that printing equipment had become entirely concerned with things I knew about—0s and 1s—I simply *had* to develop TEX. Why? Because I knew typography was an important computer science prob-

lem. . . . And because I knew that my unique background—having worked in a print shop and with galley proofs, having found a new method for breaking paragraphs into lines—made me more likely to solve the problem well than anybody else. Really, I felt it essentially a *moral obligation* much more than a simple intellectual challenge. Curiosity had led me to develop an expertise that later I felt must be applied when I learned that I could make a special contribution. Dissatisfaction with the existing state was absolutely crucial *combined* with the knowledge that I had a special background that could help. [all emphases in the original][38]

Thus, for Knuth, the problem of digitized typography became "his" problem, partly because of dissatisfaction with the then-observed print quality—he felt that it could be improved—but, more significantly, because of his conviction that his particular background and expertise made him "more likely to solve the problem well than anybody else." It became a personal challenge, a gauntlet which he felt morally obligated to pick up.

There is a striking resemblance between Knuth's circumstances and those, recounted earlier, that prompted Alexander Graham Bell to embark on the problem of the "speaking telegraph." This further reinforces the sense that, at least in the realm of technology, curiosity, the personal drive to take up a challenge, and dissatisfaction are inextricably intertwined.

Problems Become Goals

Need, dissatisfaction, and curiosity are *mental states*. Furthermore, they are states of mind that pertain to or concern other things—a need *for* a bridge to span the Menai Strait; a dissatisfaction *with* the quality of German steel; a curiosity *about* digitized type. Philosophers call mental states that are about other things, including objective entities in the physical world, *intentional* states. They have pondered deeply on the "problem of intentionality," that is, how mental states can ever be about other things.[39]

For the present purposes, we can take the existence of such intentional states for granted. There is ample evidence, as the examples cited illustrate, that these states of mind—need and dissatisfaction in particular—are what give rise to technological problems. Furthermore, regardless of whether a problem is an expression of need, dissatisfaction, or curiosity, it is translated into a *goal*. The general form of a technological goal may be stated as

Build (or design or develop) an artifact such that it satisfies a set of requirements.

Goal making, then, is the first stage common to all technological creation. Without a goal it is inconceivable that the act of invention can ever begin.

But it is one thing to have identified or framed a goal; it is quite another to produce an entity—an artifact that meets the goal. There is an enormous gap between the goal Maurice Wilkes identified—

Develop a control unit form that is easy to design, maintain, and repair.

—and the elegant architecture of the microprogrammed control unit (figure 5-1) which Wilkes made public in 1951. Likewise, the tubular Britannia Bridge (figure 2-1) seems a far cry from the goal posed to Robert Stephenson:

Design and build a railway bridge to span the Menai Strait at the site known as the Britannia Rock so as to be sufficiently strong and stable and also satisfy the navigational requirements imposed by the Admiralty.

The possibilities whereby each of these goals could be satisfied seem endless. The "space" of possible solutions is vast. At the same time, the technologist may not, in many cases, have even a sense as to the form of the solution itself—that is, what form the artifact will take.

So, what kind of cognitive features can we postulate for a technologist that may allow us to demonstrate how a set of goals for an artifact that has never existed before can actually lead to a particular artifact that apparently satisfies the goals? To answer this, we must reflect first on *the technologist as a cognitive being*. Let us, then, turn to this issue.

4

The Technologist as a Cognitive Agent

In an article called "The Architecture of Complexity," published in 1962, Herbert Simon articulated the nature of complexity as it is manifested in both the natural and the artificial worlds.[1] A system, according to Simon, is deemed complex if it is composed of a large number of parts or components that interact in non-obvious ways. In such systems, even if one knows the properties of the components, it is a far from trivial matter to infer the properties of the whole. In a complex system, the whole is indeed more than the sum of its parts.

Simon's notion of what constitutes complexity is perhaps too narrow to be acceptable to all. It may not be meaningful, for example, to explain the contents of such literary works as Virginia Woolf's *The Waves* or T. S. Eliot's *The Wasteland*, or the intricate structure of a classical Indian raga in terms of "components that interact in a non-obvious way"—though, instinctively, we would not hesitate to label these works as complex. Nevertheless, the scientifically minded will find Simon's definition appealing, since it appears to capture the essence of what makes systems of the kind studied by scientists, economists, and engineers so difficult to understand.

There is, however, another feature that seems to characterize entities that we intuitively regard as complex: they can be explained or comprehended at many different levels. Complex systems are meaningfully characterizable in terms of *multiple description levels*.

As Gordon Bell and Allen Newell have pointed out, these levels are not equivalent in the sense that anything said one way can as well be said another.[2] On the contrary, each description level is *autonomous*, meaning that the structure and behavior of a relevant system can be completely specified, in some particular sense, at that level. At the same time, the description levels are *hierarchically related* in that given two such levels, the components constituting one level can be defined in terms of components constituting the other. The first level is then said to be "higher" than the second.

Many artifacts exhibit this property. Moreover, the property of multiple description levels is manifested in natural systems, which is why it is relevant to this discussion. For the cognitive system—the mind-brain entity—is arguably the most complex natural entity we are aware of; and one of the ways in which we commonly attempt to grasp this complexity for the purpose of describing, explaining, or understanding mental processes is by recognizing that it too admits of multiple description levels, each of which is appropriate for a particular kind of inquiry into cognitive processes.

To be more specific, using both empirical and theoretical criteria, some cognitive scientists have come to recognize the following broad levels of description for cognition:[3]

1. *The knowledge level*: Cognition is described or explained in terms of goals, actions, knowledge, and intentionally rational behavior.
2. *The symbol level*: Cognition is described or explained in terms of symbols, memory (in which symbols are held), symbol-transforming operations, and interpretations of those operations.
3. *The biological level*: Cognition is described or explained in terms of biological structures or structures that are abstract representations of biological systems—as in, for example, neural networks.

In his magnum opus, *Unified Theories of Cognition*, the computer and cognitive scientist Allen Newell pointed out that these levels of description relate to what he called distinct "cognitive bands" characterized by distinct "time scales." The biological level of description, for example, is pertinent to elements such as neurons and neural circuits where the time scale of neurological phenomena is in the range of 1 to 10 milliseconds.[4] The symbol level of description applies to a range of cognitive operations and elementary tasks requiring times ranging from 100 milliseconds to 10 seconds. The knowledge level is relevant to tasks typically requiring minutes or hours; even days, weeks, and months.

It seems that if we wish to understand the cognitive nature of technology, the most relevant level of description is the *knowledge level*. There are at least two reasons for this.

First, the time scale for the cognitive band at the knowledge level is precisely that at which technological creativity is most manifested. Invention and the design of new artifacts, contrary to common perception, are not events that normally happen "instantaneously," in the time scale of milliseconds; they occur over significant, that is perceptible, time spans. Technological creativity entails *deliberative* action. And deliberation takes time.

Second, technological creation, as we have seen in the preceding chapter, is fundamentally a goal-oriented activity. It entails actions that strive to meet goals. This is precisely what the knowledge level of cognitive description addresses.

The approach taken in this chapter, then, will be as follows. We shall first establish a coherent and internally consistent description of a cognitive system at the knowledge level. The technologist, being a cognitive agent, will conform to this general description. And yet, not all cognitive agents can design or invent new artifacts. Not all cognitive agents are *technologically* creative. The technologist, when entrusted with a goal, presumably embarks on a cognitive process of a particular kind. Since the intellectual heart of technology is design, let us call this special kind of cognitive act *the design process*. Our purpose in this chapter, then, is to arrive at a comprehensive knowledge-level characterization or "model" of design as a cognitive process. Our intention, of course, is to see whether and how this model lends us insight into the nature of technological invention.

Cognition at the Knowledge Level

The mind as an object of intellectual curiosity has a long and broad pedigree, as Richard Gregory has documented in his historical study *The Mind in Science*.[5] Since the Second World War, the advent and subsequent development of the digital computer has had its own profound impact on our attitude towards the mind. On the one hand, beginning with Alan Turing's celebrated paper of 1950, "Computing Machinery and Intelligence,"[6] investigations into the possibility of constructing machines that exhibit intelligence-like behavior rapidly grew into the science and technology of *artificial intelligence* (AI)—a discipline which, thirty-something years after its "official" birth (at a conference held in 1956 at Dartmouth College in the United States), continues to provoke controversy regarding what computers can and cannot do.[7]

On the other hand, the notion (also originating in the early days of electronic computing) that computation entails, in a broad sense, not the processing of numbers or "bits" (that is, ones and zeroes) but of symbol structures—of which numbers and bits are merely particular examples, as are mathematical expressions, linguistic texts, and pictures—has served to provide a powerful intellectual tool and *metaphor* with which psychological processes can be probed. The relatively young discipline of *cognitive science*—which brings together cognitive psychology, artificial intelligence, linguistics, neurophysiology, and other kindred disciplines, is thus the most visible intellectual marker of the influence of computers in the study of the mind. If AI aims to imbue machines with intelligence, then cognitive science hopes to imbue the mind with machine-like features or characterize it as an information-processing device.[8]

Much of what might be called classical artificial intelligence—"classical" because, of late, an alternative paradigm called "connectionism," rooted at the biological level of description has made an appearance[9]—is based on concepts and ideas related to the knowledge level; hence the latter has always been tacitly recognized, sometimes by different names, as a distinct and significant level of description for mindlike entities. For instance, the philosopher Daniel Dennett, in 1978, wrote of the "intentional stance"—this being essentially a reference to cognitive features at the knowledge level—while the cognitive scientist Zenon Pylyshyn referred to this level in a 1980 paper by such adjectives as "mentalistic" and "representational."[10] However, the term *knowledge level* was actually coined by Allen Newell, who in a 1982 article presented a systematic and detailed characterization of cognitive behavior at this level.[11]

Let us call a cognitive system at the knowledge level an *agent*. Newell, in fact, ascribed the knowledge level to both natural (that is, cognitive) and artificial (or computational) systems. An agent, in his formulation, can be as much a computer as a person. For the purposes of this discussion, however, we are *solely* interested in (human) cognition. The main entities with which an agent is concerned are goals, actions, and knowledge.

Goals

We have already encountered, briefly, technological goals in the previous chapter. In general, a *goal* is a directive to the agent—which may have originated either from an external source (through some social mechanism) or internally, in the agent's mind—to *do* or *achieve* something. For example a systems engineer, in the course of designing a high performance computer system, may be posed with the following goal:

> Create a file system such that (a) the programmer interface to it is powerful, compatible with other standard interfaces, and will hide the architectural details of the underlying hardware; (b) it provides uniform and guaranteed response to the user irrespective of the computational workload; (c) it provides a high level of performance; and (d) to the applications programmer it appears as a single hierarchical file system.

This goal, in fact, is quite complicated—being composed of several subgoals. It is also rather vague—notice such qualitative and fuzzy terms as "powerful" and "high level of performance." Generally speaking, goals can be imprecise and quite complex in scope, as in the above example, or they may be very specific and unequivocal. For instance, the requirement of a civil engineer designing a wide-span structural beam might be:

> Determine the maximum strength of the beam [of some given dimension and material].

Knowledge

The term *knowledge* here embraces somewhat more than what philosophers have traditionally meant by it—namely, "justified true beliefs." Generally speaking, an agent's *knowledge body* will include facts, beliefs, rules, theories, laws, hypotheses, metaphysical suppositions, values, and particular instances of things. We shall find it convenient here to refer to the individual elements of a knowledge body as its *tokens*.

While it is really not necessary, at least at this point, to make too precise a distinction between the various types of tokens, it is useful to exemplify the general varieties of knowledge an agent may possess and also to distinguish between declarative tokens signifying what might be thought of as passive concepts and others that allow actions to be performed or inferences to be drawn by the agent.

Kepler's three *laws* of planetary motion—

1. *law of elliptical paths*: Planets move in elliptical paths with the sun at one focus of the ellipse.
2. *law of equal areas*: During a given time interval a line from a planet to the sun sweeps out an equal area anywhere along its elliptical path.
3. *harmonic law*: If T is the period of any chosen planet and R the mean orbital radius of that planet, then $T^2 = kR^3$ when k is a constant having the same value for all planets.

—are examples of declarative tokens that are contained in every college physics student's knowledge body. And, before Antoine Lavoisier wrought the chemical revolution in the late eighteenth century, all chemists subscribed to what were then thought to be facts but which we now know were false *beliefs*, namely

1. Only compounds with phlogiston in them burn.
2. When ores are heated, the phlogiston from the burning charcoal combines with them to produce metal.
3. Calcination results from metals losing their phlogiston.
4. Respiration causes phlogiston to be removed from the body into the air.

These were all aspects of the "phlogiston *theory*" proposed earlier in that century by George Stahl. To take another example, Maurice Wilkes's knowledge body at the time he was ruminating on the best way to design control units for computers would definitely have had the following particular *fact*:

> In the MIT Whirlwind computer, each arithmetic operation (with the exception of multiplication) involves exactly eight pulses issued from a control unit implemented in the form of a diode matrix.

This refers to the computer then being built at the Massachusetts Institute of Technology which Wilkes happened to have seen when he visited the United States in 1950.[12]

Scientists and (as we shall see later) technologists, in the course of their deliberations, create or postulate *hypotheses*—declarative propositions that may or may not turn out to be true. Like laws, theories, and facts, hypotheses may also enter an agent's knowledge body and be retained there as hypotheses until further corroborative evidence causes them to be accepted with more confidence—and even be accorded the status of facts; or the evidence may result in the hypotheses being refuted and eliminated from the agent's knowledge body or ascribed the status of false beliefs.

Thus, for example, as I shall further elaborate in Chapter 6, in the course of designing the Britannia Bridge, Robert Stephenson and William Fairbairn appear to have formulated a hypothesis that

> Circular and elliptical tube sections are superior to rectangular sections in their resistance to distortion and lateral pressure due to wind forces.

It turned out that tests subsequently led to the refutation of this hypothesis and its replacement by other hypotheses.

In contrast to these examples, *rules* enable mental actions or operations to be performed, such as the drawing of an inference. It is reasonable to claim, for instance, that anyone having even the most superficial acquaintance with electrical phenomena will contain in her knowledge body the rule that

> **IF**   a signal is required to pass through
>   two devices
> **THEN**   connect the devices in series.

This illustrates the way in which rules are commonly represented by artificial-intelligence researchers and cognitive scientists.[13] The general notation for depicting a rule is

> **IF**   condition
> **THEN**   operation

which asserts that if the state of affairs in a particular "world" of which an agent has knowledge is such that "condition" is satisfied then the corresponding "operation" may be performed. The "condition" itself may be a conjunction of several subconditions.

Rules are obviously very useful as ways of representing heuristic knowledge. The word *heuristic* in its original sense signifies the means that allow one to "discover." It is also used to refer to a rule of thumb or a procedure involving trial and error. In recent years, beginning with George Polya's classic book *How to Solve It*[14] and further developed by computer scientists, *heuristic knowledge* refers to any empirical or experientially based strategy, rule, or principle that may enable a problem solver to converge rapidly and efficaciously to a solution to some given problem.

Heuristics are particularly useful in the absence or paucity of "deep"—that is, theoretically sound—knowledge about a particular problem-solving arena. The history of *technological knowledge* is, in fact, largely a history of heuristic knowledge, at least until very modern times.[15]

Vitruvius's *De Architectura*[16] encapsulates and documents a vast collection of rules concerning buildings and the building of buildings. These were presumably part of the common knowledge body of all builders of his time. In the case of temples of the Doric style, for instance, he wrote:

> Let the front of a Doric temple at the place where the columns are put up be divided, if it is to be tetrastyle, into twenty-seven parts. . . . One of these parts will be the module; and the module, once fixed, all the parts of the work are adjusted by means of calculation based upon it. . . . such will be the scheme established for diastyle buildings. But if the building is to be systyle . . . let the front of the temple, if tetrastyle, be divided into nineteen and a half parts. One of these parts will form the module in accordance with which the adjustments are to be made as above described.[17]

Here, "systyle" and "diastyle" refer to different modes of column spacing while "tetrastyle" signifies a particular number of columns. Clearly, what we have here is a prescription for action, a prescription for design. As rules, they can be expressed in the following form:

> IF   column spacing is to be diastyle
> *and*
> column number is tetrastyle
> THEN   compute a module as
> width of temple / 27.

> IF   column spacing is to be systyle
> *and*
> column number is tetrastyle
> THEN   compute a module as
> width of temple / 19.5.

In the two thousand years since Vitruvius architects have, no doubt, acquired a more theoretical foundation to their avocation, one built upon deep understanding of acoustics, energy distribution, structural behavior, mathematics, psychology, and so on.[18] And yet, the modern architect's knowledge body still contains vast amounts of knowledge not unlike the rules stipulated by Vitruvius. For example, followers of Christopher Alexander will seek to employ what Alexander called "patterns"—these being nothing but heuristic principles.[19] Thus, in *A New Theory of Urban Design*, Alexander and his collaborators present seven fundamental "rules of growth" whereby the planned growth of an urban area may be guided. One of these rules:

Every building must create a coherent and well-shaped public space next to it.

This rule is further elaborated in terms of several subrules:

a. Each time a building increment is built, it is shaped and placed in such a way that it creates a well-shaped pedestrian space.
b. The building volume of the increment is itself also simple and beautifully shaped.
c. At intervals between the buildings, there are gardens. These are also carefully shaped and follow the general rule for positive space.
d. As each new building is built, the roads nearby are extended incrementally, to give vehicular access to that building.
e. Parking space is the last element in the hierarchy, and must also be placed so that buildings surround it, and its effect on the environment is reduced as far as possible.[20]

As a final and rather beautiful example of the kind of heuristic knowledge that has historically informed the technologist's mind, consider the following passage from *De Re Metallica*, Georgius Agricola's sixteenth-century classic on mining and metallurgy.[21] This passage pertains to assaying—that is, the determination, by physical and chemical means, of the composition and purity of ores.

> The colour of the fumes which the ore emits after being placed in a hot shovel or an iron plate indicates what flux is needed in addition to the lead for the purpose of either assaying or smelting.[22] If the fumes have a purple tint, it is best of all, and the ore does not generally require any flux whatever. If the fumes are blue, there should be added cakes melted out of pyrites or other cupriferous rocks; if yellow, litharge and sulphur should be added; if red, glass-galls and salt; if the fumes are black, melted salt or iron slag, litharge and white lime rock. If they are white, sulphur and iron which is eaten with rust; if they are white and green patches, iron slag and sand obtained from stores which easily melt; if the middle part of the fumes are yellow and thick, but outer parts green, the same sand and iron slag.[23]

When a prescription for action is particularly complex, it becomes elaborated and held in an agent's knowledge body in the form of a *method* or *procedure*. Computer algorithms are methods of a very precise kind, as they are intended for computer implementation. As another example, Agricola has recorded how one may determine whether pyrite ores contain gold or not by pursuing the following method:

> We determine in the following way, before it is melted in the muffle furnace, whether pyrites contain gold or not: if after three times roasted and three times quenched in sharp vinegar it has not broken nor changed its colour there is gold in it. The vinegar by which it is quenched should be mixed with salt that is put in it and frequently stirred and dissolved for three days.[24]

Alternatively:

> Nor is gold lacking in . . . [pyrites] whose concentrate from washing, when heated in the fire, easily melt, giving forth little smell and remaining bright; such concentrates are heated in the fire in a hollowed piece of charcoal covered over with another charcoal.[25]

Some four hundred years later, the development of one of the strangest and most modern of artifacts, software, produced a widely used token of technological knowledge, the principle of "structured programming," also called "stepwise refinement." This was first proposed as an explicit doctrine in the early 1970s by the computer scientists Edsger Dijkstra and Nicklaus Wirth.[26] The principle can be stated in the following manner:

1. If the problem is so simple that its solution can be obviously expressed in a few lines of programming language, then the problem is solved.
2. Otherwise decompose the problem into well-specified subproblems such that it can be shown that if each subproblem is solved correctly, and these are composed together in a specified manner, then the original problem will be solved correctly.
3. For each subproblem, return to step 1.

By the mid-1970s this principle of program design had become an essential token of knowledge for most students and practitioners of computer programming. Donald Knuth, who began work on his TEX typesetting software system in May 1977, wrote, "TEX . . . was my first nontrivial structured program . . . consciously applying the methodology [of] Dijkstra, Hoare, Dahl . . . I found that structured programming greatly increased my confidence in the correctness of the code while the code still existed on paper."[27]

The distinction between rules and methods is, of course, not very sharp. One might think of them as constituting operational or procedural knowledge of different degrees of complexity. Furthermore, the distinction, at least in the technological mind, between declarative knowledge, such as facts, hypotheses, laws, and theories, and procedural knowledge, as represented by rules and methods, is by no means as clear as the old epistemological dichotomy between "knowing that" and "knowing how" would have us believe. The technologist as a cognitive agent often freely transforms declarative into procedural knowledge, facts into rules.

No discussion of knowledge as an aspect of cognition can be complete without at least a brief mention of one other kind of token—the agent's *Weltanschauung*: the way an agent conceives the world. His worldview. Of particular interest to us here is the worldview to which a technologist qua technologist subscribes—what we might call his *technological worldview*.

Philosophers and historians of the natural sciences have come to recognize that any understanding of how particular discoveries are made, especially those of the seminal kind, will very often depend on one's understanding of the discoverer's worldview. An essential element of the backdrop of Copernicus's heliocentric theory was a belief in the perfect geometric harmony or order of the universe—a belief handed down from the ancient Greeks.[28] And what greater expression of order and perfectness than that the universe is spherical and that heavenly bodies should move in uniform circular motion? Such was Copernicus's worldview, or at least a key aspect of it; his cosmology, as Kuhn has remarked, revolutionary though it was in removing earth from the center stage, was deeply Aristotelian insofar as the geometric essence of heavenly motion was concerned.[29] This was the worldview that Kepler was to inherit in the early seventeenth century and which he would demolish with his celebrated laws.[30]

Similarly, Newton in his *Principia* (published in 1687) laid out four rules of reasoning that not only can serve as guidelines to, and place constraints on, the making of scientific hypotheses but also, as noted by the physicist and historian of science Gerald Holton, reflect a deep faith, on Newton's part, in the fundamental uniformity of nature.[31] These rules can be stated as follows:

1. *The principle of parsimony*: Use the smallest number of hypotheses necessary and sufficient for explaining a phenomena. (This principle is also known as "Occam's Razor" after the medieval English philosopher, William of Occam).
2. *The first principle of unity*: Similar effects have similar causes.
3. *The second principle of unity*: Properties found to hold for all bodies within the scope of observation or experiment are taken to hold (albeit tentatively) for all bodies in general.

4. *The principle of faith*: Hypotheses arrived at inductively are to be viewed
as true until they are shown explicitly to be otherwise.

In the light of our earlier discussions, it is not too fanciful to think of these
as very abstract, very "high-level" heuristics constituting key elements of
Newton's own worldview.

Among the most celebrated of scientific worldviews is that of Albert Einstein,
who to the end of his life refused to accept quantum mechanics as the ultimate
description of the microphysical world. This reluctance lay in his conviction
that the principles of causality and determination were the fundamental gover-
nors of nature; and, therefore, the indeterminacy which quantum mechanics
had introduced into the scheme of things was merely a consequence of our state
of ignorance about the ultimate nature of physical reality. "Subtle is the Lord
but malicious He is not," he is said to have remarked.[32]

In more recent times, the neurophysiologist Roger Sperry, awarded the Nobel
Prize for his work on the nature of the "split brain," has written profoundly, in
Science and Moral Priority, what in effect is a statement of his philosophical
stance on the mind-brain relationship. For Sperry, the conscious mind is an
emergent property of brain activity which in turn plays a vital causal role in the
control of brain function. Or as he put it, "mind moves matter in the brain."[33]

The worldview of the creative technologist is hardly likely to be as cosmic
in scope as those of Copernicus, Newton, Einstein, or Sperry. Moreover, if the
scientist-philosopher—the scientist who consciously reflects on and explicitly
records his worldview—is a rare breed, still rarer is the engineer-philosopher.
Nonetheless, if one examines with care the writings of the most creative of technologists, one can obtain some glimpses of the nature of their worldviews.

As an instance, the Italian civil engineer and architect Pier Luigi Nervi has
stated in some of his writings that his fundamental philosophy of design is that
a necessary condition for architectural (by which he meant aesthetic) excellence
in a building is the excellence of its structural (that is, engineering) characteristics. In other words, good architecture entails good structural design. It is worth
noting that Nervi's *Aesthetics and Technology in Building* was originally made
public as the Charles Eliot Norton lectures at Harvard—lectures usually devoted to poetry and literary criticism.[34]

Still dwelling in this realm, the civil engineer David Billington, in his detailed
study of the bridges designed by the Swiss engineer Robert Maillart, has shown
that the latter's "style" was to select a structural form from which the analysis
and calculation of the internal forces in the structure would follow naturally.
In Maillart's philosophy, the mathematical analysis of structural form was sub-
ordinated to, was a handmaiden for, the development of structural form. In
Billington's words, "force followed form."[35]

Actions

Cognitive actions are the third type of entity characterizing an agent at the
knowledge level. An action *does* something. The effect of a physical action of

the kind we are familiar with is to change the state of the external world in some way: a line is drawn, a sentence written, an object moved, and so on. A knowledge-level action, in contrast, changes the state of the agent's knowledge body: a new belief is generated or a hypothesis formed, an old belief is discarded, a sentence is framed, a goal is identified, and so on.

To understand the nature of an action more clearly, consider an example. We imagine an experienced computer-systems engineer in the midst of designing a "filing system"—the component responsible for storing information in and retrieving it from, data files—for a new high-performance computer system. At some point in this process she is faced with the following *goal*:

G0   The filing system must be such that a high input/output (I/O) bandwidth is obtained.[36]

It is reasonable to assume that the engineer's knowledge body will contain all sorts of tokens, some of which pertain to systems and how one may assess their performance. Suppose that among these tokens are the following *rules*:

R1   **IF** there is a goal to achieve a high bandwidth for any data-transmission system **THEN** establish as a subgoal the achievement of a high *sustained* bandwidth.

R2   **IF** there is a goal to achieve a high bandwidth for any data-transmission system **THEN** establish as a subgoal the achievement of a high *maximum* bandwidth.

In other words, the engineer knows that a high bandwidth for data transmission may be obtained either by attempting to obtain a high "sustained" (or average) bandwidth (over some working life period of the system) or a high "maximum" (or peak) bandwidth.

Let us then suppose that the goal G0 causes the retrieval, by some form of association, rules R1 and R2 from the agent's knowledge body. The agent is *reminded* of these rules through association with the goal.

The agent's knowledge body will also contain, we suppose, tokens having nothing to do *per se* with computer systems but that pertain to general problem-solving methods, procedures, and reasoning rules. One such rule may be the following:

R3   **IF** a goal (or a fact or a rule) *A* is the case under consideration
and
there is a rule "**IF** *A* is the case under consideration **THEN** the goal (or fact or rule) *B* is the case"
THEN establish *B* as the goal (or fact or rule).

This is, in fact, an *inference rule*—that is, a rule that allows an agent to make an inference providing that the premises of the rule are satisfied. Logicians call this particular rule *modus ponens*. If we know that the presence of smoke indicates a fire nearby and we happen to observe or smell smoke, then we infer that there is a fire nearby.

Rule $R3$ itself may be retrieved or accessed by association with the combination of goal $G0$ and rule $R1$, or the combination of $G0$ and rule $R2$—or by the two combinations, one following the other. In the first case, if $R3$ is then actually applied by the agent, a new goal results:

> G1　The filing system must be such that a high sustained I/O bandwidth is achieved.

In the second case the application of $R3$ by the agent will also result in a new goal:

> G2　The filing system must be such that a high maximum I/O bandwidth is achieved.

Thus, starting with the $G0$, the agent has been led to identify $G1$ and $G2$ as subgoals—in the sense that if $G1$ is met then it will have contributed to $G0$ being satisfied, and if $G2$ is met then this will also contribute to $G0$ being achieved. The sequences of events producing $G1$ and $G2$, respectively, constitute two actions, and we can schematically depict them in the manner shown in Figures 4-1 and 4-2. I shall refer to the components of these actions, namely "retrieve" and "apply," as *subactions*.

These examples illustrate many of the general characteristics of a knowledge-level action. An action has an *input* to it—in the examples, the goal $G0$. It does something and in so doing, it produces an *output*—these being, in the examples, goals $G1$ and $G2$. The input and output, in the case of knowledge-level actions, are both *symbol structures*—they are structured sets of symbols. In our examples, the symbol structures are both English texts, but this is a matter of convenience; at the knowledge level of cognition we are not particularly concerned with *how* symbol structures are represented or encoded; that is a problem of cognition at lower levels of description, possibly the symbol but more

Input	G0: a goal about filing systems and high I/O bandwidth
	*
Retrieve	R1: a rule about high bandwidth and its relationship to high sustained bandwidth
Retrieve	R3: the rule of modus ponens
Apply	R3: with G0 and R1 as substitutions in the condition and operation parts of R3
	*
Output	G1: a goal about filing systems and high sustained I/O bandwith

Fig. 4-1. *A knowledge-level action* that begins with an input goal, retrieves by spread of activation two rules, and produces a goal as output.

Input	G0: a goal about filing systems and high I/O bandwidth
	*
Retrieve	R2: a rule about high bandwidth and its relationship to high maximum bandwidth
Retrieve	R3: the rule of modus ponens
Apply	R3: with G0 and R2 as substitutions in the condition and operation parts of R3
	*
Output	G2: a goal about filing systems and high maximum I/O bandwidth

Fig. 4-2. *Another knowledge-level action.*

likely the biological level.[37] It is sufficient for us to assume or stipulate that at the knowledge level, the inputs and outputs are represented by symbol structures; and, furthermore, that these symbol structures are, really, *symbolic*: They designate entities (and relationships between or among entities) in some universe, and therefore such symbol structures can be interpreted (and carry meaning) with respect to that universe. For instance, the text shown as G0 refers to and designates such entities in the universe of computer systems as filing systems, input/output, and bandwidth.

Goals are not the only symbol structures of interest. The rules and other tokens contained in the agent's body are also symbol structures. In fact, actions may themselves be composed into and represented by symbol structures, as Figures 4-1 and 4-2 suggest. More generally, we shall claim that all goals, knowledge, and action pertaining to an agent are representable, at the knowledge level, by symbol structures.

An action consumes some amount of *time*. We are not particularly interested in the actual duration of an action, for it may vary widely, ranging from seconds to minutes or even hours. It is reasonable to expect that actions such as those depicted in Figures 4-1 and 4-2 would consume, in the case of an experienced (or, for that matter, theoretically knowledgeable) engineer no more than a few seconds. The same goal when encountered by the novice engineer or by a student with little prior background in computer-systems design may necessitate several minutes, if not hours, to generate goals G1 and G2.

Generally, though, we must recognize that an action has a beginning and an endpoint in time. Consequently, one action may begin (or end) earlier (or later) than another. This means that two or more actions can occur in strict sequence or they could be overlapped in time—be performed in parallel. In the case of our examples, I have not actually stipulated the order of the two actions A1 and A2. It is possible for them to be performed sequentially—that is, A1 could be performed to completion before A2 begins (or vice versa); however, since the two actions do not actually interfere with or depend on each other, in the sense that the output of one does not serve as the input of the other, they could as well have taken place in parallel. Generally speaking, we shall refer to any combination of sequential and parallel actions as a *structured* set of actions.

Rational and Nonrational Actions

In the case of our fictitious systems engineer brooding over the design of the filing system, the actions shown in Figures 4-1 and 4-2 were both linked with goals—they are depicted as being invoked in response to goals. In Allen Newell's theory of the knowledge level, this kind of behavior on the part of the agent—that is, the invocation of an action in response to a goal—is a hallmark of rationality: The agent is fundamentally guided by what Newell called

> *The principle of rationality*: If an agent has knowledge that one of its actions will lead to one of its goals, then the agent will select that action.[38]

Indeed, given the original goal $G0$ and that the engineer's knowledge body is assumed to contain the rules $R1$, $R2$, and $R3$, the performance of actions $A1$ and $A2$ appear to satisfy this principle. For, they produce two subgoals, both of which, if met, will lead to the attainment of the original goal. The actions $A1$ and $A2$ insofar as they were responses to goals are, thus, instances of *rational* actions.

However, it is quite conceivable for an agent to perform a knowledge-level action even in the absence of a specific goal. Indeed, the identification or discovery of a problem in science or technology, so often the hallmark of the creative mind, may entail a knowledge-level action that produces a problem (a goal) as output where such an action is invoked not by a prior goal but simply by the agent's attention being focused, by chance or otherwise, on a phenomenon or fact. For Maurice Wilkes, the inventor of microprogramming, his observation—

> Existing control units are difficult to maintain, repair, and design

—was transformed into a problem, a goal:

> Develop a control unit form that is easy to maintain, repair, and design.

In *Creativity in Invention and Design*, a cognitive study of the invention of microprogramming, I have described a plausible sequence of knowledge-level actions whereby Wilkes might have made the mental transition from an observation (or, more precisely, a privately held opinion) to a purposeful goal. These actions all have the same general characteristics as those shown in Figures 4-1 and 4-2. Each has one or more inputs that cause, as subactions, other tokens in the agent's knowledge body to be accessed associatively and retrieved; and the rules to be applied result in inferences that generate as outputs other tokens (including goals). Their specific details are not important here. What is relevant is that none of these actions required a goal as input.[39]

In general, then, an action may be initiated in response to a goal or by virtue of a token other than a goal. I shall, therefore, distinguish as a matter of convenience between *rational actions*—those that are goal-stimulated—and *nonrational actions*—actions that are initiated by the stimulus of some entity other than a goal; for instance, a fact, a rule, or any other kind of knowledge token.

Bounded Rationality

Appearances can be deceptive. Suppose we were able to observe (by whatever means) an agent choosing or performing an action in response to a goal. According to Newell's principle of rationality, we should infer that the agent does, in fact, possess the requisite knowledge that the action so performed will lead to the desired goal.

But this inference on our part may be wholly unwarranted. For instance, the agent's knowledge may be *erroneous*. Henry Petroski has chronicled how Vitruvius in *De Architectura* and Galileo in his *Dialogue Concerning Two New Sciences*, separated though they are by seventeen centuries, both discuss the erroneous assumption that "scaling up" (that is, increasing the size of) a structure from scale model to actual size conserves the properties observed in the former. An engineer who designs or analyzes a structure on the basis of this assumption—that is, takes a particular action or sequence of actions because *his* knowledge body contains this assumption as a token—may think that he is acting rationally. He will scarcely seem to be rational to us however, especially if the structure fails because of the incorrect assumption—as has happened with many ships, airships, bridges, and buildings even in this century, almost two millennia after Vitruvius and some three centuries after Galileo.[40]

There is also the possibility that the knowledge which an agent draws upon as a basis for action is *incomplete*. One may know the truth but not the whole truth. One can hardly ever hope to know the whole truth. And yet, decisions must be made; actions must be taken based upon whatever knowledge one has at one's disposal.

We do this in day-to-day life, as much in making up our minds whether to take an umbrella when leaving the house as in deciding whether or when to sell some stocks we may own. It is also, however, a part of the landscape of more deliberative actions of the kind we like to think pervades technology.

The principle of rationality already seems more an ideal than an approximation of how an agent behaves. But there is more to follow. The realm of problem solving in many an area, whether it is technological invention, scientific discovery, or management decision making, is what is often called *ill-structured*, a concept first described by the psychologist Walter Reitman and further analyzed by Herbert Simon.[41] What this term means is that if we imagine such kinds of problem solving as entailing a series of mental actions or decisions, each pertaining to a different aspect of the problem or to different stages of the problem-solving process, then the "space" of possible actions or decisions is potentially unbounded.

Consider the design of something as pedestrian as the teakettle. For each of its features several design choices exist—Should the spout be angularly shaped or a smooth curve? What material should be used for the body? Where precisely should the handle be attached? What should it be made of? Should the kettle be filled through the spout or through a separate opening to be covered by a lid and, if the latter, how should the lid be positioned?—decisions or choices all of which are driven and constrained by the requirements of functionality, appearance, cost, safety, and manufacturability. However, there is the further complication that a particular solution for one feature or requirement precludes certain solutions for others or may even demand the exclusion of some other requirements: A handle attached to the end and curving forward toward the spout but not connected to the front is aesthetically pleasing but not likely to satisfy the functional demand, on the part of some consumers, that the kettle should be easy to pour from; such a design might also interfere with the ease with which the lid can be opened. Thus, additional decisions are entailed as to how such *conflicts* should be resolved.

If the space of solutions facing the designer of teakettles seems formidable— and the dismal fact that kettles and other mundane artifacts continually fail in their intended purposes attests to the nontrivialness of such technological problems[42]—one can appreciate how much larger is the space of choices, actions, and decisions for such artifacts as computers, spacecrafts, and cities.

Thus, given a goal related to an ill-structured domain, the cognitive agent may not be afforded the luxury of behaving according to the principle of rationality. Even if possessing all the knowledge in the world about the different disparate elements pertaining to the goal—all possible materials to use for the kettle, all possible shapes for the spout and so on—to understand fully their interactions and material dependencies, to generate the knowledge and choose to act accordingly as to approximate the principle of rationality, may be far beyond the agent's cognitive capacity and resources.

In general, then, even if an agent knows what to do in order to choose an action to perform, there may not simply be enough time (or other resources) to determine the action. The principle of rationality is thus best regarded as a behavioral ideal for the knowledge-level agent. This is how we fondly imagine an agent *ought* to behave in the world of goals, knowledge, and action.

In 1946, Herbert Simon published his *Administrative Behavior*. Based on his empirical studies of how people in organizations actually make decisions, Simon suggested that human decision-making occurs under conditions of limited, or what he later termed *bounded* rationality.[43] Almost thirty years later Simon was awarded the Nobel Prize in economics for this and subsequent work on the theory of bounded rationality, an award which no doubt puzzled all those economists who think of their discipline and profession as belonging in the pristine realm of axioms and theorems rather than that of psychology.[44]

Simon's insight, in essence, can be stated as

The principle of bounded rationality: Given a goal, an agent may not possess correct or complete knowledge of, or be able to economically compute, the correct action (or sequence of actions) that will lead to the attainment of the goal.

Presented this way, the principle seems disarmingly simple and quite commonsensical. Yet, its consequences are far-reaching for the light it sheds on all aspects of goal-oriented behavior—especially, as we shall see later, on the nature of the processes of invention and design.

Accessing Knowledge by Spreading Activation

So far, nothing has been said of how tokens are actively accessed in a knowledge body. In Figure 4-1, for instance, rules $R1$ and $R3$ are simply asserted to have been "retrieved" in the performance of the action. In fact, the knowledge-level theory of cognition is deliberately reticent about both how knowledge is represented or organized and its mode of access. All that is conceded about the knowledge body is that it *exists*, is *unbounded*, and that its constituent tokens *are accessible*.

This "abstracting away" of some obviously important details is one of the main attractions of the knowledge level, as it permits us to concentrate on the knowledge itself as the source of an agent's behavior. Issues such as where and how knowledge is held or represented, how it is accessed, or the time for its access enter the cognitive picture at the lower levels of the description hierarchy.[45]

Nonetheless, the character of the actions depicted in Figures 4-1 and 4-2 seems to suggest that, as far as the knowledge level is concerned, the general nature of access to tokens is association based. In the case of Figure 4-1, $G0$ is a goal about "filing systems" and "high I/O bandwidths"; this led, by association with the "high bandwidth" concept, to the retrieval of $R1$, a rule about "high bandwidth"; $R1$, in conjunction with the original goal, caused rule $R3$ to be accessed, again by association, since what $G0$ and $R1$ are about, match the condition part of $R3$. A similar sort of process is tacitly evident in the case of the action shown in Figure 4-2.

The assumption that, insofar as the knowledge level is concerned, tokens are accessed by association is itself founded on the principle of what cognitive psychologists call *spreading activation*.[46] The essence of this principle relies on the notion that an agent's knowledge is organized in the form of a *network* in which entities such as rules, facts, concepts, and so on are represented by the "nodes" of the network while the "links" between nodes signify certain kinds of connections or relationships between the entities represented by the nodes.[47]

Given a knowledge network, activating a node, say by association with a goal, results in the activation of adjacent or associated nodes. These, in turn, activate other nodes adjacent to *them* and so on. Thus, activation *spreads* through the network, though the amount of spread decreases over time and distance from the original node. A knowledge token is activated to the extent it is related to the source of activation.

The Dynamics of Knowledge Bodies

Assuming that the notion of knowledge organized as a network is acceptable as a useful model, the reader will no doubt recognize that associative links, can be of *greater or lesser strengths*. That is, the connections between nodes in an agent's knowledge body can differ in strength. Furthermore, links connecting the same entities may be stronger in the case of one agent than in another. The city of Chicago, to take an arbitrary example, has a variety of associations for most people. To some it is inextricably linked (rightly or wrongly) with crime and gangsterdom. For others such an association may be overshadowed by other connections. For the architect, Chicago is linked irrevocably with Louis Sullivan and the Chicago school. For the obsessive basketball fan, Chicago *is* the Bulls and Michael Jordan. But for the sports fan who also happens to be a historically minded physicist or a historian of science, a possibly *stronger* association may well be with the names of Enrico Fermi and Subramaniam Chandrasekhar, or with the atomic pile—the first ever nuclear reactor, built in Chicago.

Knowledge is a dynamic, shifting entity. We may also, thus, assume that the strength of a link in a network can *vary over time*. Connections that are strong today may become, through disuse, weak tomorrow and then, because of shifts in attention, strengthen once more the day after. As an undergraduate student of metallurgy, the concept of steel was linked in my mind with a host of other metallurgically relevant entities: the iron-carbon diagram, the elements nickel and chromium (these being constituents of stainless steel), the Bessemer converter, rolling mills, the English city of Sheffield, and so on. Since becoming a computer scientist I have no doubt, over the years, encountered steel in many different contexts. And yet, while Sheffield has remained strongly associated in my mind with steel, until recently I would have been hard-pressed to recall when it was that I last thought about the iron-carbon diagram. The link must have become very weak for I had quite forgotten about it—a fact I realized recently when reading an encyclopedia article on steel. The link has evidently strengthened since, for *now* I can hardly avoid remembering the iron-carbon diagram when I think of steel.

Links between entities can become so strong that, together with the nodes they connect, they may coalesce into a single composite entity—and, thus, form a network node. A rule such as

IF   the goal is to achieve an acceptable performance level for a system
            and
            the performance level is quantitatively unspecified
            and
            there are no specified optimization criteria for performance

THEN establish as a subgoal the achievement of a performance level for the system that is better than the performance level of its predecessor or its existing competitors.

is an instance of a composite token of the kind an engineer may have been taught explicitly or has learned through experience. This rule embodies a strong link between the properties "acceptable performance level" and "performance level quantitatively unspecified" on the one hand and the system property "better performance than competitors or predecessors" on the other.

The forming of a simple composite entity, such as a rule, by assembling previously known tokens and integrating them into a single unit is an important means whereby cognitive agents learn. The process itself is referred to as *chunking*.[48] Composite entities may, of course, be considerably more complex than rules. To take our earlier example, the concept "steel" for the metallurgist embodies an entire network of nodes and links that pertain to a variety of things: its chemical composition, its "microstructure," its common varieties, certain locations where it is manufactured, its manufacturing process, its mechanical properties, and so on. Steel for the structural engineer, on the other hand, is constituted of a network that only very weakly overlaps the metallurgist's network; for the engineer, steel is a structural material and the concept primarily embodies its mechanical and structural properties, its maintainability, or its character when present in reinforced and prestressed concrete. In common language, the metallurgist and the structural engineer have essentially "different things in mind" when they think of steel.

In psychology, such a complex of strongly linked entities is said to constitute one's "mental set" and, indeed, the much overused term *culture shock* refers essentially to the situation when one's mental set concerning ways of living does not match what one encounters in a particular culture. Other terms have also been used to denote this same notion. For example the computer scientist and AI pioneer Marvin Minsky, in the context of representing knowledge in computers, referred to such composite entities as *frames*[49]—a concept that drew upon the British psychologist Frederick Bartlett's much earlier notion of *schemata*.[50]

The reader is to be reminded that such issues as I have been discussing in the last two sections, vital though they are for understanding cognition, belong properly to the symbol and biological levels. At the knowledge level, these structures and mechanisms are not visible. The "phenomenology" of the knowledge body at the knowledge level is comprised solely of the tokens of knowledge themselves, the notion that tokens are always accessible, and the idea that tokens, once admitted into an agent's knowledge body, are there indefinitely—although certain *properties* associated with a token may change over time. Thus, the physicist who before 1900 believed in the continuous nature of energy would, even after the publication of Max Planck's famous law, retain the token or concept "continuous nature of energy" in his knowledge body; however, its status would in all probability have been revised and changed from "fact" or "strongly held belief" to "false" or "myth."[51]

Finally, one should also note that at the knowledge level, there is no explicit structure to the knowledge body. If when focusing on a concept I am reminded immediately of another concept, I may, at best, *infer* that there is a direct and strong link between the two. If this remembering takes an appreciably longer time, I may infer that either the link is weak or that the two concepts are at a "distance" from one another in the knowledge network and, thus, several intermediate links had to be traversed. Which of these, if any, is actually the case is irrelevant at the knowledge level.

Knowledge Level Processes: Rational and Nonrational

The picture that has emerged so far of the nature of cognitive actions can be summarized in the following terms.

1. Every action has, as input, one or more symbol structures, representing goals and/or knowledge tokens.
2. The action is deemed rational if at least one of its inputs is a goal—that is, if the action is goal driven. Otherwise, it is said to be a nonrational action.
3. Every action produces one or more symbol structures as output—these also denoting goals and/or knowledge tokens.
4. Every action entails the retrieval and application of tokens contained in the agent's knowledge body. Such retrieval occurs by association and the spread of activation through the knowledge body.
5. In performing a rational action, the agent is driven by the principle of rationality; that is, if she believes or knows that a particular action will lead to the achievement of the goal, then she will select or perform that action.
6. However, the agent is cognitively constrained by bounded rationality— an action performed may not be the correct one insofar as the goal is concerned; or the agent may not be able to economically determine which action to take.
7. Nonrational actions, not being goal oriented, are not driven by the rationality principle. Nor are they subject to bounded rationality.
8. Every action consumes time.

We have also taken note that actions can be performed in sequence or in parallel and that, in the most general situation, combinations of sequential or parallel actions may be performed. I have referred to such general combinations as constituting a "structured" set.

Now, any structured set of operations, events or actions taking place over time, whether in nature or in the world of artifacts, is normally called a "process"; thus, accordingly, we are led to the notion of a *knowledge-level process*, which is, simply, a structured set of knowledge-level actions.

It is useful, especially given that our interest in this book is on technological thinking, to make the distinction between *rational* and *nonrational* processes.

Here, I am clearly drawing an analogy with the two types of actions distinguished earlier. A rational knowledge-level process is a knowledge-level process that has, as an input to its first or earliest action(s), at least one goal. The process as a whole is initiated in response to one or more goals. In contrast, a nonrational process is one that is not initiated by a goal.

The dividing line between the two kinds of processes is, on occasion, somewhat fine. I am driving to work one morning and as I negotiate the traffic my mind dwells on a dream I had the night before. Like many dreams, this seems utterly strange when recollected in daytime for it contained totally unrelated and incompatible places, people, and events that had been welded into a scenario which in my dream state had seemed entirely coherent. As I ponder this, I suddenly realize that there is a remarkable similarity between the bringing together of unrelated or incompatible elements in a dream and the concept of *bisociation*. This was the novelist Arthur Koestler's term for the combining of elements from two or more entirely unrelated domains—or, in Koestler's words, "matrices of experience." To Koestler, this was the essence of the act of creation.[52] My noting of this similarity between the combining of incoherent elements into a logical scenario in my dream and Koestler's notion of bisociation leads me to hypothesize that "dreams are acts of creativity that occur in the sleeping state."

Clearly, a knowledge-level process has just been described. Beginning with a "fact"—the occurrence of a particular dream—I was led to invoke a knowledge token representing Koestler's theory of bisociation, and by means of some sort of reasoning I was led to a specific symbolic output, a hypothesis connecting dreams and creative acts. However, so far as this scenario is concerned, the input to the process was not a goal; it was simply the fact of the dream itself. The process is, thus, an instance of a nonrational process, although its outcome is an interesting hypothesis worthy of future investigation.

However, had I initially asked myself the question "What kind of explanations can be postulated for this dream?" a knowledge-level process very similar to the one just sketched above and leading to the same output could well have occurred, the main difference being that the process would be invoked by a problem or goal. The latter process would be of the rational kind. We may conclude that the first scenario led to a hypothesis "by chance," serendipitously, while the second produced the hypothesis through purposeful means. Both, however, are knowledge-level processes—explicable in terms of knowledge and actions.

Design as a Knowledge-Level Process

The model of cognition described above, if valid, is obviously as applicable to technologists as to other human beings. However, not all knowledge-level processes are acts of technological creation. Not all cognitive agents are technologists. Nor do all cognitive agents have the capability for designing or inventing artifacts. A cognitive process relevant to technology is likely to be a special instance of the more general process.

Let us then turn to *design*, the intellectual epicenter of technological creation. As we saw in Chapter 2, design as a separate and separable element of the technological process is by no means inevitable. The craftsmen in history have rarely made the distinction between conceptualization—design—and making. Even now, notwithstanding the protestations of the "progressives" in software engineering, many computer programmers still deign to make the distinction. For them, designing and building software are entirely inseparable.

Nonetheless, design as an act is central to all the varieties of technology, and since the creation of artifactual form, including what we conventionally call invention, is so patently and irrevocably associated with the act of design, we can surely hope to gain significant insight into the nature of the inventing mind by focusing on the design process itself. If we can even begin to understand design as a cognitive process, then we will surely have made some headway in understanding the technological mind.

One of the very real problems one encounters in characterizing the act of design is to produce a definition that on the one hand satisfies our intuitive idea of design and on the other permits useful and interesting conclusions to be extracted from the definition. William Addis began his book on the nature of the relationship between theory and design in structural engineering by listing the various different meanings of *design* given in *The Oxford English Dictionary* along with the dates of their earliest recorded usage.[53] *To design* is, thus, to plan out, trace the outline of, delineate, draw, plan and execute, make the preliminary sketch of, make the plans and drawings necessary for the construction of, fashion with artistic skill; *a design* is a plan or scheme conceived in the mind of something to be done, the preliminary conception of an idea to be acted upon, a preliminary sketch for a work of art, the combination of details which go to make up a work of art, and so on.[54]

Those who are professionally engaged in design—the creators of artifacts—attempt, on occasion, to formulate more systematic definitions. Thus, for Christopher Alexander, design is "the process of inventing physical things which display new physical order, organization, form in response to function."[55] Others, including David Pye and this writer, recognizing the difficulty of identifying what exactly *is* function and also the desire for other goals of design (e.g., reliability and performance), have eschewed altogether the idea of a one-sentence definition. Rather, they prefer to think of design as a network of features and characteristics, a frame or a schema in the sense of artificial intelligence.[56]

Nonetheless, given that design is, after all, a cognitive act, it behooves us to try to place it within the framework of the knowledge-level theory of cognition.

Let us propose that a *design process* is a knowledge-level process that is subject to a number of constraints.

First, the input to the process includes symbol structures that designate or represent a set of properties to be met by some artifact in some given universe. These properties constitute a *set of requirements*. The requirements will certainly pertain to function—the artifact must be able to do such and such—but it may also stipulate other desirable properties of the artifact: performance, for example—the artifact must be able to do what it is intended to do within a given time or using no more than a certain amount of some resource; or reliability—the artifact must function for such and such time before failing; or aesthetics; or cost.

Second, the output of the process is a symbol structure that represents the artifact itself. This representation we will call *the design*. A set of engineering drawings, a list of instructions, a handbook describing the structure of an organization, an algorithm or a computer program, architectural sketches and drawings, even a set of equations describing the behavior and structure of the artifact—all are instances of such output. They are all symbol structures, some graphic or pictorial, others textual.

Third, the goal of the agent in conducting the process is to produce a design D, say, such that if an artifact is built according to D it will satisfy the properties constituting the requirements R. We will refer to this as the *design goal*, and for convenience we can state it concisely as "D satisfies R." The design goal is also an input to the process. Thus, the design process is goal driven. It is a *rational* knowledge-level process.

Finally, the agent's knowledge body should contain no token corresponding to any other artifact or design that satisfies the requirements. That is, a design process is initiated only when the agent is unaware of any other artifactual form satisfying the given set of requirements. For if the agent already knows of such an artifact, there would be no need to initiate design. According to the principle of rationality, the agent would simply produce the known solution. Thus, "newness" (even in the most modest sense) is a condition needed for a design process to occur. *One designs in order to initiate change.*[57]

Notice, though, that newness is always relative to the agent's knowledge body. For example, given a set of requirements for an integrated-circuit chip, the experienced engineer, one whose knowledge body contains many "cases" of prior designs—Thomas Kuhn would call them the exemplars that are part of one's paradigm[58]—may know of a chip design that already satisfies the requirements. There is, then, no need to design a chip. The relevant action here is to retrieve the prior design. The same problem, given to a student or an engineering trainee would most likely initiate a design process since the agent might have no knowledge of appropriate cases.

These two scenarios pinpoint a well-known distinction made by cognitive scientists and AI researchers between *deep* and *compiled* knowledge, or between *preparation* and *deliberation*.[59] The experienced engineer in the example resorts to prepared or compiled knowledge. The neophyte must deliberate, access more-basic knowledge, and build up a solution.

Most real design situations fall between these two extremes. For instance, the civil engineer may be posed with requirements for a highway bridge that includes details of the required span, the local soil conditions, the topography of the region, and expected loads. These requirements may be found to be similar but not identical to the characteristics of some other bridge-design problem known to the engineer, either through personal experience or the technical literature. In that case, a design process will be initiated that in all probability will take the known design as a starting point. The process may even entail modification of that design. On the other hand, it may turn out that some crucial aspect of the requirements is absolutely novel so far as the engineer's knowledge body is concerned. In that case, he may have to resort to deep rather than compiled knowledge. On occasion, as in the case of Robert Stephenson and the Britannia Bridge, this might even lead to an entirely new bridge form.

Radical or Inventive Design

This last point reminds us that though design as a process is initiated in order to bring about a change—and, thus, that design is fundamentally an act of creative cognition—the extent of creativity will vary considerably from one situation to another. Not all technological acts lead to patents or new artifactual forms. Most engineers, planners, architects, and other technologists probably spend the greater part of their lives with design problems that do not prompt them to challenge the fundamental technological paradigm within which they operate. Every now and then, however, a new artifactual form is established, or a new approach to the creation of artifacts is proposed. Such technological events we may legitimately call *invention*. On occasion, inventions may actually result in the older designs for that particular class of artifacts being gradually abandoned as design choices and replaced by the newly invented form; or, at the very least, a new viable alternative becomes available to the technologist. In effect, a new paradigm is established.

This scenario is reminiscent of Kuhn's portrayal of how revolutions in science can occur.[60] In technology as in science there is the possibility of both "normal" and "revolutionary" activities. The aeronautical engineer Walter Vincenti refers to this as the contrast between *normal* and *radical* design.[61] Elsewhere, I have referred to the latter as *inventive* design.[62]

It is, thus, of obvious interest to examine more clearly the cognitive nature of invention or radical design, wherein the highest form of technological creativity is exhibited. This becomes our next topic.

5

The Connection between
Invention and Design

In his book *The Nature and Aesthetics of Design*, David Pye sought to distinguish between invention and design. According to him, invention is the discovery of a *general principle of arrangement* that, in effect, governs or defines a class of systems. Design, in contrast, is the *application* of such a principle in some given context and results in a particular embodiment of the principle.[1]

That there is some distinction to be made between the two concepts seems plausible enough. We often do tend to use the words in different ways and senses. And when we think beyond technology, there are, undoubtedly, contexts in which employing the word *invention* seems perfectly valid but *design* does not. Newton and Leibniz are both credited with the invention of the infinitesimal calculus in the seventeenth century. We do not say that they designed the calculus. Likewise, Darwin and Wallace are said to be the coinventors of the theory of evolution by natural selection—meaning both that they proposed an explanation of how the evolution of species may have come about and that they discovered the natural principle of the law of evolution. It is clearly quite absurd to state that they designed a theory. Or for that matter, that they discovered a theory.

In the realm of technology however—and this was the domain which he had in mind—Pye's particular distinction between invention and design demands closer examination. There is no doubt that there are technological instances that appear to meet Pye's characterization of invention. In 1855–56, Sir Henry Bessemer took out his first patents for the manufacture of steel by the process known subsequently as the Bessemer process.[2] We call his invention the Bessemer process because, as described then and thereafter in the technical literature, it is stipulated as a general set of operating and procedural principles distinct to that process. Similarly, von Neumann, Eckert, Mauchly, and their collaborators are said to be the inventors, in 1946, of the stored-program computer, although they were not the first to actually design and build a physical computer. Their invention was constituted of the detailed organization and principle of the stored-program computer form as described in a famous widely circulated but never published memorandum written by von Neumann.[3] In contrast, Maurice Wilkes and his colleagues in Cambridge *designed* and built the EDSAC, just as Frederick

Williams and his colleagues in Manchester *designed* and constructed the MARK I. Both these machines, completed in 1949, were the first concrete embodiments of *the* stored-program computer. Wilkes, however, *invented* the principle of the microprogrammed control unit in 1951. His scheme was, indeed, a principle of arrangement (Figure 5-1). Later, he and his Cambridge collaborators *designed* the EDSAC 2. Its control unit was a concrete realization of his invention.[4]

We can, however, offer counterpoints to these examples. Mauchly and Eckert in Philadelphia and John Atanasoff and Clifford Berry in Ames, Iowa, both laid claim to having invented the modern electronic computer because the particular computers they happened to have designed and constructed—ENIAC and the Atanasoff-Berry computer respectively—were historically original.[5] Thomas Newcomen's invention of the atmospheric engine entailed his conceiving (that is, designing) and constructing the first working atmospheric steam engine in 1712, used to operate a mine-drainage pump in Staffordshire, England.[6] Alexander Graham Bell's invention of the telephone is more precisely marked by his design and construction in June 1875 of a particular instrument, the gallows telephone.[7] In all these cases, invention clearly hinged on designs of specific artifacts.

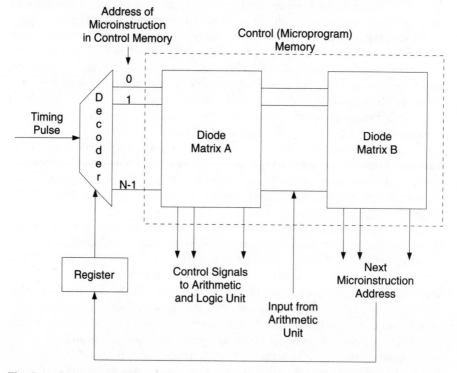

Fig. 5-1. *Organization of the microprogrammed control unit as invented by Maurice Wilkes in 1951.* The control unit is responsible for issuing control signals that enable and ensure the proper electronic events to occur in a computer in the proper sequence. Prior to Wilkes's invention, control unit circuits were largely irregular and ad hoc in structure. Wilkes's achievement was to produce a form for the control unit that was vastly more organized and regular.

It seems, then, that the connection between invention and design is some-what more diffuse than Pye would have us believe. Let us, then, probe more closely the nature of the inventing act.

The Test of Inventionhood

Regardless of whether we are referring to technology or some other realm, the litmus test of invention as a process is that it results in an *artificial product* that is *original* in some significant sense. There are, then, two criteria for a process to be deemed an invention.

First, the product must not be a feature of, or belong to the world of nature. James Watson and Francis Crick did not invent the double-helical structure of the DNA molecule. That structure has existed since practically the beginning of life on earth. They discovered the nature of this structure. Nor did Roentgen invent X-radiation. He discovered it in the same sense as Columbus is said to have discovered America. And while we do talk of Boyle's law, it was not *his* law in the sense that the atmospheric engine was Newcomen's or the telephone was Bell's. Boyle's law is a natural law characterizing the behavior of gases which Robert Boyle happened to have discovered.

The second criterion for inventionhood is that the artificial product must be original in some significant way. This must be the case whether the product is utilitarian—and thus an artifact in the sense I described in Chapter two—or otherwise. Benjamin Huntsman is credited with the invention of the crucible process of making steel because the resulting process yielded a steel of *higher quality* than was previously available. Alexander Graham Bell invented the telephone in the sense that this device constituted an entirely *new mode of communication*. In the mathematical domain, Newton and Leibniz are said to have co-invented the calculus because the calculus was an absolutely and significantly new kind of technique of mathematical analysis. Darwin and Wallace simultaneously invented the theory of evolution by natural selection in that their theory embodied a new explanation of a biological phenomenon. Notice that theories *about* nature, like artifacts, are invented. Laws *of* nature are discovered.

Even though we may grant that originality is a critical and necessary hall-mark of inventionhood, we may debate as to what it means to be original "in some significant sense." How precisely can we pin down the notion of origi-nality? How original must the product of one's thought be to be deemed an invention? Is originality a personal, subjective matter, or is it determinable objectively or by society at large?

Originality and Creativity

Originality and creativity are obviously connected very intimately, and several writers on creativity have addressed questions such as these. In her book *The Creative Mind*, the cognitive scientist Margaret Boden proposed the distinction between what she called "psychological creativity" (or P-creativity) and "historical creativity" (or H-creativity).[8] When someone has an idea that is fundamentally novel for the individual who had the idea it is said to be P-creative. An idea or thought that is novel with respect to the whole of history is said to be H-creative.

Thus, for Boden, creativity is a property of the product of one's mind and, for all intents and purposes, it is synonymous with "fundamental" novelty. In my book *Creativity in Invention and Design* I suggested instead that creativity is a feature not of the product of one's thinking but of the agent's cognitive act or process that resulted in the thought product. And, as a further refinement of Boden's distinction between the psychological and the historical, I proposed characterizations of the concepts of originality and creativity along the following lines.[9]

We have already seen in Chapter 4 that central to what constitutes a cognitive agent at the knowledge level is the notion of a knowledge body. Let us refer to a knowledge body, associated with a particular agent and pertaining to a particular domain, as the agent's *personal* knowledge body (about the domain). This is to be contrasted with a knowledge body associated with, and shared among, a community of agents and pertaining to the same domain. This latter we may refer to as the community's *public* knowledge body (about the domain).

There is, of course, nothing sacrosanct about the states of such knowledge: What constitutes an agent's personal or a community's public knowledge body concerning a particular domain may change over time. Today's fact may well become tomorrow's fiction. Or what was thought yesterday as the state of the art is today relegated to the realm of an aside in the history of the topic. A paradigm shift, in Kuhn's classic sense, exemplifies how an entire scientific community can transfer its allegiance from one set of "truths" or beliefs to another.[10]

Consider now, a state of affairs in which a particular knowledge-level *process* conducted by an agent results in a thought product π relevant to a particular domain D. In general, π may be an idea, a mathematical theorem, a scientific hypothesis, a design for some artifact, a method or procedure, or some such like. D, of course, is the field or discipline to which the thought product belongs. Sometimes, D may be the combination of several fields. Thus, for example, Johannes Kepler's third law of planetary motion, $T^2 = kR^3$, relating the periodic or orbital time T of a planet to the mean orbital radius R (where k is a constant for all planets), is a thought product that belongs to both physics and astronomy. Donald Knuth's computerized typesetting system TEX is an artifactual thought product belonging to a domain consisting of printing technology, algorithms, and programming technology.[11]

We will say that π is *psychologically novel* for the agent if he believes that there exists no other thought product in his personal knowledge body that is identical to π. If, furthermore, the agent also believes that π adds significantly to the relevant community's *public* knowledge body associated with the domain, then we shall say that π is *psychologically original* for the agent.

If on the other hand (or in addition), the relevant *community* believes that there is no other existing thought product in its public knowledge body that is identical to π, then π will be said to be *historically novel* for that community. If, in addition, the community agrees that π adds significantly to its knowledge body then we shall say that π is *historically original*.

It will be noticed that the notions "is identical to" and "adds significantly to" are not defined or specified in any precise sense whatsoever. Whether some thought product is identical to another, or whether it adds significantly to knowledge, is determined by the standards, criteria, and values held, respectively, by the individual agent and the relevant community, and will themselves be constituents of the relevant respective knowledge bodies. I shall return to this matter shortly.

We can now distinguish between different *levels* of creativity as follows:[12] A process conducted by an agent giving rise to some thought product π relevant to a domain will be said to be *PN-creative* if π is psychologically novel for the agent; it will be said to be *PO-creative* if π is psychologically original for the agent. The process will be deemed *HN-creative* if it is considered by the relevant community to be historically novel. Finally, it will be called *HO-creative* if π is historically original for that community.

Perhaps the most unsettling implication of these definitions is that there appears to be nothing special about the knowledge level process itself that demarcates creative from noncreative processes or, for that matter, that distinguishes between different levels of creativity. Whether or not one's efforts in generating some thought product are deemed creative is determined *after* the process is over and the thought product itself is available for assessment. Likewise, whether or not such a process is considered to be PN-, PO-, HN-, or HO-creative is also determined after it is done with. Moreover, there is always the possibility that a thought product considered at first by the community to have been historically novel is, in the light of later knowledge, acknowledged to be historically original. Or that something thought by the agent to be psychologically original is found by him later to have already been discovered or invented—in which case, it must be "downgraded" to the psychologically novel category.

There is perhaps no better example of the presence of these different categories of originality in a single body of work than that of the Indian mathematician Srinivasa Ramanujan (1887–1920). The remarkable history and circumstances of Ramanujan's life are by now well known owing in considerable part to the written testimony of the eminent British mathematician, G. H. Hardy, who, so to speak, discovered Ramanujan and retrieved him from comparative obscurity.[13] A recent biography by Robert Kanigel has painted a more complete picture of Ramanujan's life and work (and, indeed, of Hardy's too).[14]

The story, very briefly, is as follows. In 1913, while still a totally obscure clerk in the southern Indian city of Madras, Ramanujan, who had been unable to pass the most basic of college examinations, sent a bundle of mathematical manuscripts to Hardy, then a Fellow of Trinity College, Cambridge, with an accompanying letter. Hardy's immediate reaction to these manuscripts has been vividly described, with his novelist's panache, by C. P. Snow. Some of them were already known to Hardy but most of them looked quite fantastic. Irritated, he put the papers aside and went about his day's work.[15] And yet, throughout the day, he could not help thinking about the manuscripts. He had never seen or imagined such theorems before.[16]

That same evening, after dinner, Hardy and his distinguished collaborator, J. E. Littlewood, examined the manuscripts in detail. By midnight, as Snow described it, they were certain that these manuscripts were the work of a man of genius.[17] A quarter of a century after this fateful episode, Hardy himself was to comment:

> Ramanujan's letters to me, which are reprinted in full in the *Papers*, contain the bare statement of about 120 theorems, mostly formal identities extracted from his notebooks. I quote fifteen [here] which are fairly representative. They include two theorems . . . which are as interesting as any but of which one is false and the other, as stated, misleading. The rest have all been verified since by somebody. . . .
>
> . . . I have proved things rather like (1.7) myself and seem vaguely familiar with (1.8). Actually (1.8) . . . is a formula of Laplace first proved properly by Jacobi; and (1.9) occurs as a paper published by Rogers in 1907. . . .
>
> . . . [The formulas] (1.10)–(1.12) defeated me completely; I had never seen anything in the least like them before. . . . They must be true because if they were not true, no one could have had the imagination to invent them.[18]

About two-thirds of this early work by Ramanujan, Hardy then went on to say, had been anticipated by others.[19] Thus, a substantial number of the theorems Ramanujan had sent to Hardy in 1913 were, clearly, psychologically original for the Indian mathematician working, as he was, in isolation from the mainstream of Western mathematical thinking. But they were not historically original nor, indeed, even historically novel. Some of the other theorems proved to be simply wrong, and while from Ramanujan's own perspective they would probably count as psychologically original, we can, at best, call them psychologically novel since they were, apparently, not contained in Ramanujan's knowledge body of the time. But the remainder, those that Hardy "had never seen anything in the least like . . . before," were certainly, according to Hardy's account, historically novel and many historically original. And, as Kanigel's biography has documented, a large part of his published work, both before he made contact with Hardy and after as Hardy's collaborator and colleague at Trinity College, were unequivocally deemed by the mathematical community to be historically original.

The Criteria of Historical Originality

This brings us back to the issue of the criteria by which a thought product is judged to have added significantly to the prevailing knowledge body. In particular, how is it possible to establish that a scientific result, a piece of mathematical work, or a technological achievement is historically original?

This question in the case of the natural sciences is closely related to the issue of how scientists reach consensus or how disagreements are settled in the matter of a scientific problem; and philosophers of science, both past and present, have been much exercised by this issue.[20] It would be pleasant to report that such inquires have yielded elegant, formal, objective tests to determine whether or not some theory is historically original. There does not appear to be any such test—and it is unlikely that any will be forthcoming, given that even in the relatively well-structured domain of science, something accepted and thought of as historically original today may well be unceremoniously rejected in the future. Science, like all thinking that pertains to the "real" world, is what logicians call *nonmonotonic*—that is, what is believed to be true at a particular time is a function of the knowledge and evidence at hand at the time.[21] Later knowledge or evidence may cause us to *revise* our beliefs. The phlogiston theory was widely held to be valid until Antoine Lavoisier demonstrated otherwise in the 1770s; for over two hundred years, the physical world was believed to behave according to Newtonian mechanics until Max Planck in 1900 and Einstein in 1905 together showed the contrary.

But even though there may exist no objective tests of historical originality, there are several historical and social criteria that can and are used for adjudicating the worth of scientific and technological works.

To begin with, any thought product in the natural sciences or in technology that leads (immediately or eventually) to a paradigm shift or to a major change in the conceptual structure of knowledge in that field is usually judged to be a highly significant contribution to the relevant knowledge body. Such work is, indeed, deemed revolutionary in the event that it alters the very belief structure that the relevant community (or some major part of the community) holds about its domain. The Darwinian revolution effected by Charles Darwin and Alfred Russell Wallace; the chemical revolution brought about almost singlehandedly by Lavoisier; the Copernican, Newtonian, relativistic, and quantum revolutions in physics are the classic examples from the natural sciences.[22] In the technological domain, the invention of fire-making almost a million years ago, of the wheel in the fourth millennium B.C., and of the weight-driven clock in about the fourteenth century fall very much in the same revolutionary category.[23] Most economic and technological historians would also place Gutenberg's invention of printing with movable type circa 1450, the Wright brothers' development of the first powered flying machine in 1903, and the birth of the stored-program computer in the 1940s (to take but a few obvious examples) at the same general level of significance.

However, for one of the more systematic and thoughtful descriptions of the character of technological revolutions, we are indebted to the structural engineer William Addis. In his study of the relationship between theory and design— that is, between engineering science and engineering design—Addis used Kuhn's theory of scientific revolutions as a framework (an intellectual paradigm in its own right) within which he identified and analyzed several major and minor revolutions in the realm of structural engineering.[24] One of these is what he termed the "Greek design revolution," begun in about the seventh century B.C. and precipitated by a "rapid transition from small scale and relatively temporary folk architecture to a monumental architecture."[25] It is the technical substance of this revolution in building-design procedures and techniques during the Hellenic period (circa 800–323 B.C.) that Vitruvius was to record much later in his *De Architectura*.[26]

Addis's second example is the "Gothic design revolution," initiated circa 1140 with the building of the Chartres and St. Denis cathedrals in France, spread over a span of four centuries, and marked by the distinctive and highly recognizable Gothic style. Finally, in modern times Addis refers to the "plastic design revolution." This occurred between 1930 and the mid-1950s, and is characterized by structural design methods for steel-framed buildings that took account of the plastic or postelastic behavior of steel.

Even if not quite bringing about a revolution—there have been, after all, only a handful of episodes in the history of science that historians, scientists, and philosophers have honored with the appellation "revolution"—the natural scientist or mathematician who makes a discovery or publishes a theoretical idea causing a new field of study to be revealed or a new class of research problems to be uncovered is usually recognized in the relevant community as having made a significant addition to the pertinent knowledge body. Such work finds its way into the textbooks (indeed, becomes the subject of whole textbooks) and review articles, and usually allows one to call the cognitive process that gave rise to it historically original. Ramanujan's work is a brilliant case in point, for his papers and his notebooks have served, and continue to serve even now, as a rich source of problems for the mathematical community.[27] G. H. Hardy himself, after his protégé's untimely death at the age of thirty-three, would continue to be influenced by the latter's stream of ideas. In the words of Robert Kanigel, Hardy's papers were "fairly littered with Ramanujan's name: 'Note on Ramanujan's Trigonometrical Function $Cq(n)$ and Certain Series of Arithmetical Functions' appeared in 1921; 'A Chapter from Ramanujan's Notebook' in 1923; 'Some Formulae of Ramanujan' in 1924. Then more in the mid-1930s: 'A Formula of Ramanujan in the Theory of Primes;' 'A Further Note on Ramanujan's Arithmetical Function $\tau(n)$,'"[28] And in 1936, Hardy would deliver a series of lectures at Harvard University that formed the basis of his later book *Ramanujan: Twelve Lectures on Subjects Suggested by His Life and Work*.[29]

A similar example is Alan Turing's seminal work on computability. In a celebrated paper published in 1936, Turing (1912–1954)—in his own way, as legendary a figure in modern mathematical lore as Ramanujan—had set out to address a question posed in 1928 by the great German mathematician David Hilbert. Namely: is there a definite method that, in principle, is guaranteed to correctly demonstrate whether or not a mathematical assertion is true? Mathematicians and logicians (and, in more modern times, computer scientists) refer to this as the decidability problem. In the course of addressing this question, Turing invented a purely symbolic, abstract machine now known as the Turing machine which, he showed, has the property that any thing that intuitively counts as being computable can actually be calculated by means of a sequence of operations of his "machine," that is, by a purely mechanical process.[30]

Turing's work set the stage for what much later came to be known as automata theory or the theory of computation. In fact, it may be legitimately argued that insofar as Turing's 1936 paper heralded the creation of a new paradigm, it was revolutionary in character.

It is interesting to compare, in passing, Turing's 1936 work on computability, in the performance of which he was unquestionably HO-creative, with another instance of his creative capability, described by Margaret Boden.[31] In 1934 Turing submitted a fellowship dissertation to King's College, Cambridge. It turned out that, unknown to him, his results had been anticipated by a Scandinavian mathematician. According to Turing's own knowledge body, this work was psychologically original. It was not, however, historically original. Nonetheless, acknowledging his personal (that is, psychological) creativity in rediscovering an important result, King's College awarded him the fellowship. This was an explicit recognition of his PO-creativity.[32]

In the realm of technology, it is also possible to identify criteria by which one may judge something to have contributed significantly to the relevant public knowledge body, even if it falls short of being revolutionary in the sense that fire, the wheel, and the movable type were. Just as in science and mathematics, where ideas and discoveries are deemed significant when they spawn new fields of inquiry or research problems, so also in technology, artifactual forms or designs are acknowledged to be significant if they give rise to copies of the original or modified versions of the original or to an entire class of artifacts rooted in that original version. Imitation is not only the sincerest form of flattery; it is a hallmark of technological originality. If in science a significant idea or a discovery unveils a new field of inquiry, in technology a significant idea or artifact or design opens up, as it were, an industry.

Maurice Wilkes's invention of microprogramming falls very much in this category. Prior to Wilkes, the few control units that were in existence were ad hoc, irregularly structured circuits designed in a particularly unsystematic way. Such control units were not only difficult to design correctly but were hard to maintain and repair. Wilkes proposed a highly regular organization for the control unit that would be designed systematically and was much easier to maintain and repair (Figure 5-1).[33] This, then, was Wilkes's contribution to the public knowledge body of the time pertinent to computer design. That this contribution was historically *novel* at the time it appeared is a matter of empirical fact. That it was also judged historically *original* is supported by subsequent events. The first computer to incorporate the microprogramming principle in the design of its control unit was the EDSAC 2, built by Wilkes and his colleagues in Cambridge between 1952 and 1958. Between the mid-1950s and the early 1960s, several variations on the original design were published by scientists and engineers in the United States, Europe, Japan, and the Soviet Union. In the mid-1960s, the microprogramming idea was adopted commercially by IBM and used extensively and very successfully in their influential System/360 series of computers. This, in turn, paved the way for widespread adoption of this type of control unit by other manufacturers.[34] In 1967, the first of a series of international conferences on microprogramming was organized, this series continuing to be held to this day.[35] And, in 1970, the first textbook on the subject was published.[36]

It is also worth examining, in the same light, another of my running examples, the case of Benjamin Huntsman and his invention of the crucible process in the 1740s. Huntsman has a very secure place in the history of iron and steel making in particular and the history of technology in general. Yet, if we are to acknowledge the "age of steel" as one of the principle technological ages of civilization—if we are to acknowledge, that is, that the invention and development of steel was in the nature of a technological revolution—then this "age" is usually identified, both by metallurgists and historians, with the first patents taken out by Henry Bessemer in England in 1854–55 and by William Kelly in the United States in 1857. Bessemer's and Kelly's patents on the pneumatic process marked the beginning of what is usually called modern steelmaking. Huntsman's development of crucible steel occurred a full century earlier, and, thus, seems to be regarded more as part of the prehistory of steelmaking. As an undergraduate student of metallurgy, for example, my studies of steelmaking began, properly, with the Bessemer process.

Yet, to repeat, Huntsman holds a very important place in the history of metallurgy and technology. Samuel Smiles devoted a considerable discussion of Huntsman in his *Industrial Biography: Iron Workers and Tool Makers*, first published in 1863.[37] He is mentioned in *Technics and Civilization*, Lewis Mumford's important critical study of the history of the machine.[38] The U.S. Steel Corporation's magisterial multiple edition treatise, *The Making, Shaping and Treating of Steel*, a standard reference on the metallurgy of iron and steel, devotes a not insignificant section to Huntsman's process.[39] In what sense, then, has history judged Huntsman to have been so original in respect to the invention of the crucible process? What was so significant about his contribution to the public body of metallurgical knowledge?

Huntsman, it is to be remembered (from Chapter 3), was primarily a clockmaker and toolmaker. He was a steel user and it was his dissatisfaction with the available steel of the times that prompted him to embark on his experiments in steelmaking. The best steel available at the time was "shear steel" and it was produced by a method called the cementation process. Briefly, this involved heating iron packed in charcoal in a furnace. The longer the time for which this was done, the deeper would the carbon penetrate into the iron. The process, however, was long, requiring as much as three weeks.[40] It was thus a costly process, and the better-quality steel that could be produced by this process was only used when it had to be. For most other purposes cast and wrought iron were used. But these could not be forged or rolled to the precise shapes or dimensions that were often required.

Huntsman's contribution lay in the development of an efficient process for making better-quality steel than had been produced before. It involved melting shear steel in a crucible, thereby allowing the carbon to spread uniformly throughout the molten steel, pouring out the molten metal into ingots and letting it solidify. Prior to Huntsman, steel had never been molten! This was a radically new idea and must surely constitute his most significant single contribution to metallurgical knowledge. Indeed, according to Kenneth Barraclough who has written extensively on the pre- and early history of steelmaking,

Huntsman's invention formed the basis for all subsequent ingot-making steel melting processes.[41]

That the relevant technological community ascribed high significance to his invention is, like Wilkes's much later achievement, borne out by the subsequent history. Huntsman did not patent the process but attempted to keep it a secret. However, many others sought to copy the process over the next three decades and one such attempt, so an interesting anecdote goes, involved industrial espionage wherein an ironmaker, one Samuel Walker, disguised as a beggar, sought shelter at night in Huntsman's works and was allowed to enter by workmen who took pity on him. Feigning sleep he observed the whole process leaving the following morning with his newly gleaned secret.[42] The interest, indeed excitement, that Huntsman's invention caused, especially abroad, is also evidenced by the many foreign observers who visited Huntsman's establishment near Sheffield in England and who, in turn, helped establish steelworks in their own countries as a consequence of their visits. The final and perhaps most unequivocal indication of the significance of Huntsman's invention is the fact that prior to his experiments, begun in 1740, no more than 200 tons of steel were being manufactured in the Sheffield region. A century later, Sheffield was producing some 20,000 tons a year, over 40 percent of the total produced in Europe. By 1862 the overall production of steel in Sheffield had risen to about 78,000 tons.[43] All of this steel was produced by the crucible process. We thus have ample evidence of the significance of Huntsman's crucible process—his thought product. One has no qualms in claiming that in inventing the crucible process and cast steel, Huntsman was HO-creative.

As a third and final case study of the criteria for judging historical originality consider yet another of my running examples, the Britannia Bridge. This bridge resembles Huntsman's process in that it has a prominent place in civil engineering history much as the crucible process is securely situated in metallurgical history. As I have noted in Chapter 2, the Britannia Bridge was regarded in Victorian Britain as a symbol of its imperial eminence, and even during its construction, engineers would visit the site to see its progress.[44] In this century, Stephen Timoshenko in his authoritative *History of Strength of Materials* devoted an extensive discussion to the design history of this bridge.[45] In still more recent times, it became the subject of an important and highly readable monograph by Nathan Rosenberg and Walter Vincenti,[46] and continues to be of contemporary interest to structural engineers and historians alike.[47] Clearly, the Britannia Bridge has been, and continues to be, judged historically original by the civil-engineering community. The question then arises as to what were the criteria by which it was so judged.

Perhaps the most obvious fact about the design of the Britannia Bridge is that at the time of its construction, it was historically *novel*.[48] The bridge was essentially a tubular beam of sufficient height such that trains could pass *through* it (Figure 2-1). No tubular bridge had ever been built before. But how was this historical novelty elevated to the status of historical originality?

Unlike the case of Wilkes's microprogramming principle or Huntsman's crucible process, the source of originality in the case of the bridge did not lie in this particular bridge directly influencing subsequent bridge forms. Stephenson himself went on to build four more tubular bridges, one in Britain, two in Egypt, and one in Canada; but his bridge form certainly did not spring up all over the world or, indeed, all over Britain. In fact, it was then displaced by other types of bridges, especially by various forms of the truss bridge. Furthermore, the Britannia Bridge has been severely criticized in economic and aesthetic terms.[49] Thus, the most obvious evidence of historical originality in technology, imitation, did not really prevail in this instance.

From the perspective of its influence on subsequent civil engineering, one of the most significant factors was undoubtedly the fact that this project heralded a shift from cast iron to wrought iron as the choice of structural material.[50] In this sense, a paradigm shift may be said to have taken place. However, a much more powerful source of its originality, as we shall see in more detail in a later chapter, is the *knowledge* it generated concerning the behavior and strength of wrought-iron structures at a time when there was practically no relevant theory to speak of on which Stephenson could rely. As Timoshenko noted, the design and construction of the Britannia Bridge engendered new knowledge of the strength of iron plates and riveted joints, and of the behavior of structures under lateral wind pressures.[51]

The new knowledge gained in the course of the construction of the Britannia Bridge, was also to exert influences beyond the realm of bridges. William Fairbairn, Stephenson's principal consultant in this project (whose contribution to the design of the Britannia Bridge will become evident in a later chapter), an eminent engineer, iron worker and engineering author in Victorian Britain, later introduced, very successfully, tubular sections into the construction of cranes.[52] Rosenberg and Vincenti in their detailed study of the Britannia Bridge echoed Timoshenko: They made the point that though the tubular bridge was soon replaced by other structural forms in the construction of long span railway bridges, the ultimate significance of the Britannia Bridge lay in the knowledge that was generated in the course of its design and construction.[53]

In conclusion, it would be reasonable to claim that the criterion for the historical originality of the Britannia Bridge lay partly in its innovative use of wrought iron as structural material but more importantly in its generation of engineering knowledge. The artifactual form itself, novel though it was, was from a historical point of view, far less influential.

The Act of "True" Invention

Let us then return to the proposition that the litmus test of invention in technology is that it results in an artifactual form that is "original in some significant sense." We now have a somewhat more definite knowledge-level criterion for what it would mean for an artifact or artifactual form to be "significantly" original. At the very least we can stipulate that the hallmark of what I shall call *true invention* is that it is a structured set of knowledge-level actions (that is, a knowledge-level process) the input to which is a goal to produce an artifact that must satisfy a set of requirements, and the output of which is an artifact (or a representation thereof) that is psychologically or (still better) historically original. The actions need not be only of the symbol-processing kind. They may be physical actions in the tradition of the craftsman, as for example when, given the need for a device to pick up pieces of food for the purpose of eating, some genius in ancient times fashioned the first crude fork[54] or when some enterprising blacksmith in the eighteenth century modified the traditional European trade axe to the particular form distinctive of the American axe.[55] However, insofar as much of technological creation from classical times on has involved the generation of some kind of *representation* of the artifact in the form of sketches or drawings, the actions may be entirely of a symbol-processing kind. In other words, *a design process becomes (or is) an act of true invention when its output—a design—is deemed psychologically original by the designer or historically original by the relevant community.*

Here we have, I believe, the proper connection between invention and design. True invention leads to psychologically or historically original outputs. It is thus a PO-creative or HO-creative process. Invention in the technological domain does not necessarily entail or presuppose design. Nor is every design process an act of true invention—although as I suggested in the last chapter, where the design process was defined, design is initiated with the *expectation* that the resulting artifact would be new (that is, either novel or original). However, a design process becomes an act of true invention only if the process is PO- or HO-creative.

6

Technology and Hypotheses

The principle of bounded rationality introduced earlier (see Chapter 4) tells us that there are limits to our ability to act correctly in response to goals. This view of rationality is derived from empirical considerations of how humans actually behave and act in the real world.[1] In hindsight, the principle seems utterly obvious, commonsensical, and *reasonable*, especially when we contrast it to the formal, idealized model of "perfect" rationality encountered in most discussions of economic and other forms of decision making.[2] Moreover, once we have encountered bounded rationality and have found time to absorb the idea, we discover that it informs one's understanding of a large range of goal-oriented cognitive acts. It helps shed much light on many questions pertaining to design, discovery, and invention. We also come to realize that bounded rationality is a wonderfully humbling principle, reminding us, as it does, of our cognitive limits. It makes us more understanding and tolerant of human errors.

The Hypothesis Law of Maturation

In the context of technology, we are particularly beholden to the bounded-rationality principle, for it has helped to uncover what I think is a genuinely universal law of technological reasoning, namely:

> *The hypothesis law of maturation*: A design process that reaches termination does so through one or more cycles of hypothesis creation, testing and (if necessary) modification.

The basic idea embedded in this law has long been part of the informal or "folk" theory of design. Henry Petroski, for example, has noted how structural engineering designs resemble the postulates of science. In the former case, however, the hypotheses pertain to the behavior of a bridge or a building rather than that of some aspect of the natural universe.[3]

The computer scientist and AI researcher B. Chandrasekaran, in his work on the construction of computer programs that automatically design, has used a design strategy or paradigm which he called *propose-critique-modify*.[4] That is, in Chandrasekaran's model of the design process, a design is first proposed, then "critiqued," and finally modified, thus creating a successor design proposal, and so on. This paradigm falls very much within the general scope of the hypothesis law. Another similar antecedent to the law is my concept of *design as an evolutionary process*[5]—an idea to which I will return later in the book.

The hypothesis law in the form it is stated here, however, is intended to be *testable*. It was first presented by this writer in 1992,[6] and was derived as a direct implication of the inherent bounded rationality of knowledge-level processes. I shall not repeat here the reasoning that led to the framing of the law.[7] Rather, regardless of *how* it was derived, let us examine seriously the claim made above that the hypothesis law is a genuinely universal, empirical law of technological reasoning.

By "universal," I mean that the law is valid not only across artifactual domains but across historical time. Universality of a law or principle in the context of technology implies artifact-independence and timelessness. By "empirical," I mean that the law says something about the real world. Let us consider, then, the matter of the testability of this law.

Testability of the Hypothesis Law

If we were to adhere to the doctrines of Karl Popper, the testability of an empirical proposition entails that the proposition should be falsifiable.[8] In the case of the hypothesis law, this means that we should be able to construct a test that actually aims to falsify or refute the law. And, as Popper and others have pointed out, there is no such thing as *conclusive* evidence *for* a general empirical proposition although some evidence *against* it may, in fact, turn out to be conclusive. All we can do is to critically expose the law to tests that may falsify it. If the falsification attempt fails, the test may be said to provide *corroborative* evidence for the test.

In Popper's scheme of things, falsifiable tests for a theory are devised by deducing bold predictions from that theory and subjecting such predictions to tests. Thus, falsifiability of the hypothesis law—and consequently, its testability in the strict Popperian sense—may not be possible. This is because the hypothesis law is *qualitative* in nature and, as in the case of other qualitative laws, principles, or theories, its testability does not depend on deducing testable predictions. Rather, the testability of a qualitative law must rely on the ability to gather empirical evidence of the law "at work" in the domains in which the law is claimed to hold; furthermore, the empirical evidence must be of a nature that *it may potentially fail* to corroborate the law. Testability of qualitative laws thus appears to rely not on falsifiability in the sense that Karl Popper made famous, but on the possibility of failing to corroborate. If the evidence yielded by the test supports the law, the law may be said to have been corroborated by the test and one's confidence in the law increases. If the evidence is found not to support the law, the test can be said to have failed to corroborate it, and our belief in the universality of the law is shaken. If many such failures to corroborate are found, we may conclude that the law is not universally valid.

In order to subject the hypothesis law to the test, we need to be more precise about what it means to *corroborate* the law. Let us consider a particular design process P' which is known to have culminated (that is, reached termination) with a particular output design D_n. Recall from Chapter 4 that in conducting such a design process, the designer is hoping to achieve a *design goal* G_n, that "design D_n satisfies a given set of requirements R_n." But because of bounded rationality, there is no guarantee that the process actually achieves the goal. Thus, the designer's claim that design D_n is such that G_n is met is at best a *hypothesis*—an assertion which may or may not be true. In that case, P' will serve to *corroborate the hypothesis law* if evidence can be provided that

CC1 P' actually entailed the formation of a *series of two or more hypotheses*:
        H_1: D_1 is such that G_1 is met
        H_2: D_2 is such that G_2 is met
        . . . .
        H_n: D_n is such that G_n is met

where $D_1, D_2, \ldots, D_n$ are a series of (not necessarily distinct) designs and the G_i's are a series of (not necessarily distinct) *design goals* and, furthermore, that

CC2   The transition from hypothesis H_i to its successor H_{i+1} was the outcome of the critical testing of H_i and its subsequent modification.

I shall call these the two *corroborating conditions* CC1 and CC2, respectively, for the hypothesis law. Now, if the design process with which we choose to test the hypothesis law follows a particular design method, and this method is known to have been explicitly influenced by the hypothesis law, then it can hardly serve as a critical test for the law. Our test process should in no way be biased a priori toward the hypothesis law—for then the test will scarcely have the potential to falsify the law, in the sense of failing to corroborate it. Thus, a precondition for CC1 and CC2 to be applicable is that any design process selected to test the hypothesis law was not influenced, as far as it is known, by the law in the first place.

Unfortunately, while the precondition is easily satisfied by many cases of design and invention from the past, historical episodes that can be shown to satisfy the corroborating conditions are somewhat harder to come by. Peter Medawar, the Nobel Prize–winning biologist and scientific essayist, once wrote that the scientific paper is fundamentally a fraudulent document in the sense that it misrepresents the thought process that led to the results being reported in the paper.[9] Thus, someone interested in delving into the very nature of the cognitive process underlying a particular scientific discovery may be completely misled by the stylized orderliness of the relevant paper. So also, even when one has access to papers or the patent literature pertaining to a particular design or invention, one may be entirely misled by their contents, at least as far as the nature of the underlying cognitive process is concerned.

It would be gratifying to claim that we have available extensive and exhaustive tests that both satisfy the corroborating conditions and unequivocally corroborate the hypothesis law. Such a happy state of affairs will, one hopes, be forthcoming in the not too distant future. For the present, we can expect that the search for corroborating (and potentially falsifying) evidence for the hypothesis law will remain an ongoing process. However, in the remainder of the chapter, I will first present three examples that, at least tentatively, appear to corroborate the hypothesis law.

The first example is taken from the realm of metallurgy—it has to do with the invention of a new type of alloy in the 1940s. The example is brief and corroborates only tentatively because it is taken from a single review paper written by the metallurgist R. W. Cahn.[10] This paper is, however, somewhat less of a "fraud" in Medawar's sense, for it actually attempts to give some idea of the reasoning process that went into the particular technological episode discussed.

The second short example is taken from the domain of algorithm design and is based upon a study of how one particular computer scientist developed an algorithm.

The third example is one I have alluded to several times already in this book, the circumstances that led to the design of the Britannia Bridge in the mid-nineteenth century. In this case, we are extremely fortunate that the principal participants in the project documented in great detail and with the utmost clarity their actual thinking as it unfolded in the course of this remarkable engineering project. From the perspective of testing the hypothesis law, one could hardly have hoped for better "data."

The Case of the First Superalloy for Gas-Turbine Blades

The development of the gas turbine in the 1930s brought about the need for an alloy for turbine blades that would satisfy the following set of requirements:

R_1 The alloy must be resistant to creep deformation and rupture at very high temperatures (in the range of 750–1,000°C); it must be resistant to hot gaseous corrosion and to accidental shocks; it must be capable of being formed.

The reader may recall this example from Chapter 2, where it was briefly discussed as an instance of a technological problem born of need. As also explained then, creep is the phenomenon of continuous time-dependent, nonreversible (or plastic) deformation of metals and alloys under load. Alloys that are expected to operate at high temperatures—as those used in the construction of turbine blades—are more likely to manifest or undergo creep deformation than materials operating at lower temperatures.

Cahn's account does not state any specific quantitative criteria as to the extent of creep and corrosion resistance. Thus, it must be presumed that the most desirable alloy is one that meets the above-mentioned requirements better than do others—where "better" means more resistance to creep deformation and rupture at higher service temperatures, greater resistance to gaseous corrosion, and so on. The initial design goal can thus be stated as:

G_1 An alloy that satisfies requirements R_1 better than others.

The first efforts at developing such strong alloys started around 1938 at the Research and Development Department of the Mond Nickel Company in Birmingham, England. They began to examine 80/20 nickel-chromium alloys and nickel-chromium-iron alloys. The former was well known as a material for electrical heating elements and for its oxidation-resistance properties. As for the nickel-chromium-iron alloys, their choice was apparently determined by the fact that some prior studies had indicated that these alloys were creep resistant and also amenable to a physical process called "age hardening."

This phenomenon, also known as "precipitation hardening," plays an important role in this story; very briefly, the phenomenon can be described as follows.

An alloy is basically a solid solution—that is, a mixture of two or more kinds of atoms in the solid state—in which the more abundant atomic form, the solvent, is a metal while the less abundant atomic forms, the solutes, may be metals or nonmetals (such as carbon). When pure metals are combined to form alloys, certain crystal structures can result in certain temperature and composition ranges. Each of these crystal structures constitutes what is called a distinct "phase." An alloy may be heated to a temperature at which one phase (usually present in small quantities) dissolves in the other, more abundant, phase. Suppose that the alloy is left at this temperature until a homogenous solid solution is formed and is then very rapidly cooled by immersing in a liquid coolant such as water (an operation known as "quenching"); then a supersaturated (solid) solution is obtained. The excess of the solute phase is precipitated in the matrix of the solvent phase. The most important effect of this precipitate is that the matrix is hardened when the alloy is maintained at a certain (usually above-room) temperature for a period of time. The alloy is then said to have been age hardened.[11]

Returning to the Mond metallurgists' choice of nickel-chromium-iron alloys, the fact that these alloys exhibited both creep resistance and age-hardenability suggested a possible connection between the two properties. Another previous study had related the creep strength of copper-nickel-tin alloys to their state of age hardening. However, whether these two characteristics were actually connected was then not known. It was supposed by the Mond metallurgists that, assuming there was a connection, they would have to add solutes to make the matrix of these alloys age-hardenable.

In analyzing historical cases such as this and others later in this chapter, we will find it convenient to distinguish between what I shall call "main" and "auxiliary" hypotheses. The former relate directly to the artifactual design as it evolves over time; the latter are additional relevant propositions that are advanced and tested and aid, in one way or another, in developing the overall reasoning.

In the present case, the investigation actually began, as we have just seen, with an auxiliary hypothesis:

AH_1 Age hardening is the cause of creep resistance

and prompted the following main hypothesis:

H_1 80/20 nickel-chromium and/or nickel-chromium-iron alloys suitably age-hardened could meet goal G_1.

In order to test H_1, the first tests on creep strength were performed on several commercial batches of the 80/20 nickel-chromium alloy. The tests were not very consistent and the metallurgists attributed the differences to the variation in the titanium and carbon contents of the different batches. Titanium was known to be a deoxidizer. As a result, small amounts of titanium (0.4%) and carbon (0.1%) were added, resulting in the first "superalloy," called Nimonic 75. A new hypothesis was thus formed as a successor to H_1 and as the consequence of testing H_1:

H_2 Nimonic 75: an alloy such as that G_1 is achieved.

The early search for turbine blade material did not stop with Nimonic 75. As far as one can determine from Cahn's account, the metallurgists were far from satisfied with this first superalloy. For the next step was to carry out experiments in which the titanium content was increased. The proposition under consideration was another auxiliary hypothesis:

AH_2 Increasing the titanium content of the nickel-chromium alloys improves their creep strength.

Some improvement did indeed result thereby partly corroborating AH_2. However, the tests also revealed surprising variations in creep resistance from one test batch of alloys to another. These variations were attributed to minor impurities—an assertion that quite clearly constitutes another (auxiliary) hypothesis:

AH_3 Minor impurities cause variations in the creep strength of the different batches of the tested alloy.

However, according to Cahn's account, an additional auxiliary hypothesis, called the "marginal solubility hypothesis" had been framed by the investigating metallurgists:[12]

AH_4 [Marginal solubility hypothesis] The best creep resistance is attained when the alloy contains just more of each solute than was soluble at the required service temperature.

AH_4 was proposed, apparently, because the experiments had also revealed that age hardening by itself is not *sufficient* to cause creep resistance. That is, the auxiliary hypothesis AH_1, was not true without further qualifications.

At this point, X-ray diffraction studies were carried out, and some of the X-ray patterns were such that the metallurgists were led to conclude that their alloys contained an iron-based phase known as "sigma"—a highly brittle constituent. Thus, a new auxiliary hypothesis was generated as a result of the X-ray studies (which had been conducted for the purpose of testing the hypotheses AH_3 and AH_4). We can specify this new hypothesis as

AH_5 The alloys contain an iron-based sigma phase.

It turned out, later, that the metallurgists had been mistaken about the presence of the sigma phase, and thus AH_5 was proved to be false. But at the time, this hypothesis led to the decision to use a high-purity additive; a pure nickel-titanium hardener containing some aluminum was used. This led to an improved alloy and, hence, to a new main hypothesis

H_3 A nickel-chromium-based alloy with a pure nickel-titanium hardener containing some aluminum meets goal G_1.

Thus, somewhat ironically, as Cahn pointed out,[13] a false premise—in this case, AH_5—led to a correct conclusion, or rather, a correct action.

It will be recalled that the original auxiliary hypothesis—that age hardening is the cause of creep resistance—had been proved incorrect in the sense that age hardening was not found to be a sufficient factor for creep resistance. Further X-ray studies of the phenomenon of age hardening in these alloys just produced yielded the following hypotheses:

AH_6 Age hardening in these alloys depended only on the phases Ni_3Ti and Ni_3Al.[14]

AH_7 The most effective age-hardening factor (or precipitate) was $Ni_3(Al,Ti)$ i.e., Ni_3Al or the γ' phase containing some dissolved titanium.

While it is not entirely clear from Cahn's account, the fact that these particular alloys were being analyzed for their age-hardening characteristics seems to indicate that the X-ray studies that yielded AH_6 and AH_7 were, in fact, performed to further delineate the precise scope of AH_1 in the context of the alloy mentioned in H_3. Thus, AH_6 and AH_7 should actually be regarded as being generated from both H_3 and AH_1.

The strategy suggested by the marginal-solubility hypothesis was then further applied. The composition of the variable precipitate could be controlled by adjusting the titanium/aluminum ratio while elements such as chromium, cobalt, and iron controlled the matrix properties. It was further observed by the metallurgists who had arrived at the hypotheses AH_6 and AH_7 that the matrix and the γ' precipitate had the same crystal structure, nearly the same lattice parameter, and that γ' was coherently precipitated.[15] More important, they arrived at the following conclusion:

AH_8 The coherent precipitation of γ' was essential for creep resistance.

However, they did not know why this was the case. The outcome of the above-mentioned control of the composition of the γ' precipitate was the development of both an improved nickel-chromium-based alloy and a new main hypothesis:

H_4 A nickel-chromium-based alloy with coherently precipitated γ' phase is such that G_1 is achieved.

The development of creep-resistant superalloys by no means ended at this stage, and Cahn goes on to describe further aspects of their development. Unfortunately, his continuation of the story is not sufficiently detailed for our purposes here. The brief account that I have just presented is, however, sufficient for us to ask whether in fact the process whereby the first super-alloys were invented meets the two corroborating conditions for the hypothesis law.

It seems that the condition CC1 was indeed met: the overall process entailed a sequence of (main) hypotheses, H_1, H_2, H_3, H_4. In order to determine whether or not the second corroborating condition CC2 was met, let us summarize the overall process.

Beginning with an auxiliary hypothesis AH_1, the first main hypothesis H_1 was formed. The experiments on creep strength constituted a direct test of H_1 and led to H_2, the second main hypothesis. So far so good. The next stage was the formulation of an auxiliary hypothesis AH_2 which was, in effect, the result of dissatisfaction with Nimonic 75, the alloy referred to in H_2.

In the course of testing AH_2, and also AH_1, other auxiliary hypotheses, AH_3 and AH_4 were generated. And, as a result of testing these latter hypotheses, a further auxiliary hypothesis, AH_5, was produced—this, in turn, led to the third main hypothesis, H_3. Thus H_3 was the outcome of a critical process that began with H_2, entailed the generation and testing of some auxiliary hypotheses, and ended with H_3. It appears that condition CC2 has also been met in the transition from H_2 to H_3.

In the next stage, the newly developed alloy (or, more probably, as far as one can make out from Cahn's account, a family of alloys in a compositional range) became the subject of further testing for their age-hardenability characteristics. The outcomes were, first, the auxiliary hypotheses AH_6 and AH_7 and then AH_8. The consequence of AH_8 was the fourth and final main hypothesis, H_4. The transition from H_3 to H_4 thus appears to be characterized by a process that began with a critical look at H_3. The condition CC2 once more appears to have been satisfied. We may conclude that insofar as the details are available

in Robert Cahn's survey, the development process for the first superalloys
appears to corroborate the hypothesis law.

The Case of the Convex-Hull Algorithm

If metals and alloys are among the most material of artifacts, algorithms are
among their most abstract counterparts. Alloy design belongs to physical technology. The design of algorithms can be appropriately regarded as mathematical engineering.

In 1984, Elaine Kant, an artificial intelligence researcher working with Allen
Newell, carried out empirical studies, in vitro as it were, of how technically
qualified people design algorithms.[16] The strategy used for this purpose was to
give human subjects problems that were to be solved algorithmically and to have
them talk aloud while they developed the algorithms. This vocalization of thinking is called a *protocol*. This can be recorded and subsequently analyzed.[17]

In one particular study conducted by Kant and Newell, a subject whom they
called "S2" was given the problem of designing an algorithm to find the convex
hull for any given set of points. The *convex hull* is the smallest subset of points
that when connected form a polygon containing all the other points. (Figure
6-1). The subject assigned this task had a Ph.D. in computer science and was
quite knowledgeable in algorithm design but knew little about convex hulls. The
protocol generated by S2 was tape recorded. In the course of developing the algorithm, S2 drew a number of diagrams on a blackboard which the experimenters copied. The latter would also ask occasional questions during the session.

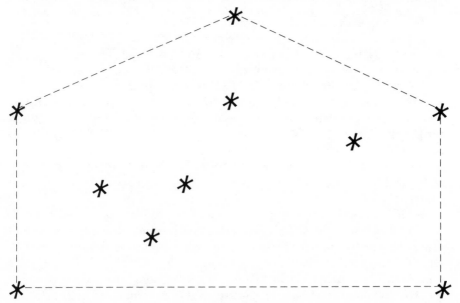

Fig. 6-1. *A set of points and its convex hull*—the smallest subset of points that when
connected form a polygon containing all the other points.

The subject actually developed two different algorithms for the problem, but here, I shall describe the development of just one of them. Based on the protocol recorded by Kant and Newell, the following distinct phases or segments in *S*2's thinking can be identified.

*S*2 was assigned the problem in terms of the following goal:

G_1 Given a set of points, design an algorithm that will generate the subset of these points that encloses all the other points.

*S*2 almost immediately responded:

Let's start with some point. Either a point is on the convex hull or it's not, right? And the question is how to make the decision.[18]

In other words, *S*2 very quickly identified an initial design of the algorithm, namely:

D_1 Select a point and test whether it is on the convex hull or not.

The algorithm would choose a point from the given set and see whether it was in the convex hull or not. If it was, the point would be in the solution— that is, be retained in the subset; otherwise, it would be discarded. We seem to have a first hypothesis here.

*S*2 also realized, as the comment above indicates, that it was not known *how* to determine whether a point selected is on the convex hull or not. This observation is clearly a criticism of the first solution. *S*2 responded to the question by next drawing on the blackboard an example diagram consisting of five points (Figure 6-2) and used this example to establish a test for determining whether a selected line—that is, a connection between two selected points—is on the convex hull or not. As the protocol recorded:

OK, let's suppose I start with a point here. And I'll just draw a line to some other point, right. Now I can go in any one of three directions from this point.

I conjecture that if it's the case that I can choose two points such that I can go on either side of the given line, then this line can't be on the convex hull. And I had better retreat. Here's *A, B, C, D, E*. So I can go from *A* to *B*. And I find that from *B* I can go either to *C* or *D*, and *C* and *D* are on different sides of the line . . . then clearly, this line can't be on the convex hull.

This test condition, which we may call "On-convex-hull-test," is itself in the nature of an auxiliary hypothesis and can be stated, almost in *S*2's words, as follows:

AH_1 [On-convex-hull-test] If it is the case that one can choose two points such that it is possible to go on either side of the given line, then this cannot be in the convex hull.

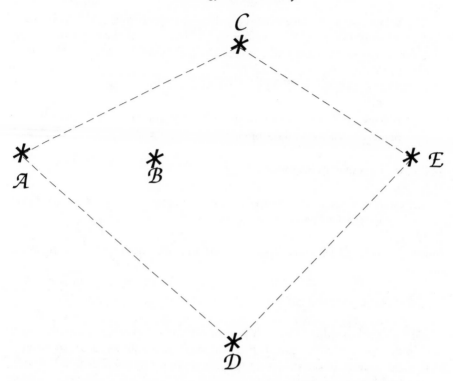

Fig. 6-2. *A test set of points used by subject S2 while deriving an algorithm to find convex hulls.* The dotted lines delineate the convex hull for this set of points.

The protocol next reveals that, using the On-convex-hull-test and the example figure as a test case, S2 elaborated the original algorithm (D_1) further. This involved, first, determining what to do if the On-convex-hull-test failed for a selected point. In that case, as S2 remarked:

> Let's retreat . . . and choose some other point.

Thus, at this stage, the algorithm was slightly modified to the following form:

D_2 Select a point and draw a line to that point from the previously selected point and apply the On-convex-hull-test. If the test holds, delete the selected point and choose another one.

The protocol's next segment reveals that S2 exposed the algorithm developed thus far to more critical scrutiny. In the subject's words:

> Let me rephrase the problem to make it even harder on myself.

S2 realized that the algorithm would not succeed by selecting *any* arbitrary point as the starting point:

> If I start at point B [see Figure 6-2] and I go to A then, either route there, I still have that problem and I want to retreat back to point B. But it turns out that

no matter what . . . which direction I go in from point *B* I'm going to have that same problem. So point *B* can't possibly be on the convex hull.

At this stage, *S2* did not immediately attempt to tackle the problem of determining how to select a "good" starting point. The protocol shows that *S2* simply chose point *A* as a known "good" starting point—

Let's go back . . . to starting at *A* because *A* is going to be on the convex hull.

—and walked through the example figure as the algorithm would do it. The algorithm thus was subtly modified to the form:

> D_3 Assuming a good starting point, select a point and draw a line to that point from the previously selected point and apply the On-convex-hull-test. If the test holds, delete the selected point and choose another one.

At this stage, the experimenters asked the subject to describe the algorithm. *S2* responded by saying:

Well, I'm not sure it's an algorithm yet, right? Because if I start at a losing point, if I were to have started at the middle point *B*, I would have found that . . . none of the segments from *B*, *B–A*, *B–C*, *B–D* or *B–E* . . . would have given me a satisfactory line. So I would have given up on *B* and tried some other . . . point to start with. So I keep doing that till I get a point that satisfies it. There must be such a point since there is a convex hull, presumably. That sounds like the algorithm.

The subject had, thus, determined a rough method for selecting a good starting point. The protocol reveals that *S2* then went on to state this more systematically and in more general terms. In fact, this method, which I shall call "Starting-point-choice-test," can be regarded as an auxiliary hypothesis:

> AH_2 [Starting-point-choice-test] Choose a point p_o, OK. Then choose a point p_1 from the remaining set of points. Draw that line segment. . . . if it's the case that there are points on both sides of that line segment then . . . give up on p_1 and try some other point . . . and keep doing that and if you exhaust all the points then p_o can't be on the convex hull. So you try another point to start with.

The algorithm was now in its final form. Figure 6-3 shows the algorithm in standard notation as it might appear in a textbook on algorithms. This constitutes the final design, D_4. Finally, *S2*, like all good algorithm designers, assessed the performance of the algorithm and concluded that it was not very good. The subject then went on to develop a second and more efficient algorithm.

The entire session leading to the algorithm of Figure 6-3 took only fifteen minutes! And, although not explicitly stated, the account just outlined makes it apparent that a chain of main hypotheses, each corresponding to a version of the algorithm, was formulated in the course of the subject's deliberations and, moreover, the transition from one version of the algorithm to its successor—

Input X is the set of points

[1] Choose a good starting point X_0 in X using Starting-point-choice-test and place it in the set "Hull-so-far."

[2] Delete X_0 from X

[3] *While* X is not empty *do*

[4]   Select a new point X_i in X;

[5]   Draw a line YX_i from the most recently added point Y in Hull-so-far-to X_i (that is, add X_i to Hull-so-far);

[6] *If* Hull-so-far is not a convex hull according to On-convex-hull-test

[7] *then* delete YX_i (that is, remove X_i from Hull-so-far)

[8] *else* delete X_i from X
        End while

Fig. 6-3. *The final convex-hull finding algorithm devised by subject S2.* Statements 1, 2, and 3 specify the major steps to be performed in sequence. Step 3 is performed iteratively—as long as the condition "X is not empty" is satisfied, steps 4, 5, and 6 are repeatedly executed in sequence. Step 6, in turn, tests for a condition; depending on whether or not this condition is met, step 7 or step 8 will be executed.

from one hypothesis to the next—was the result of testing or critically assessing the predecessor version. The protocol thus appears to reveal that both the corroborating conditions, CC1 and CC2, were met.

The Case of the Britannia Bridge

My very first "experiment" for testing the hypothesis law using historical material involved the celebrated design of the Britannia Bridge.[19] As I have already noted, this engineering enterprise provides a most remarkable case study for scrutinizing the technological process—not only for the purpose of testing the hypothesis law but from a variety of perspectives.[20] This is no doubt in large part due to the two detailed accounts written by the principal actors in the story: Edwin Clark's two-volume history of the project, published in 1850, which includes a lengthy introduction by Robert Stephenson;[21] and Sir William Fairbairn's book, published a year earlier.[22] In using the Britannia Bridge design as a test of the hypothesis law, I drew entirely on these two sources for my "data."

We have already seen from the example of the high-temperature alloys, the general nature of the historical approach: One tries to show, summoning the historical material as evidence, that the particular process of design or invention under scrutiny entailed the (possibly implicit) construction of a sequence of hypotheses bearing the general form "a particular design satisfies the given design goal," such that the two corroborating conditions CC1 and CC2 are satisfied. This same technique was, of course, used in the case of the Britannia Bridge. The evidence was comprised of first-hand accounts given by Stephenson, Clark, and Fairbairn, as well as reports of parliamentary hearings and the extensive correspondence among the main participants—especially between Stephenson and Fairbairn.

To do full justice to the "experiment" involving the Britannia Bridge, I would need to repeat practically the entire contents of the original account.[23] This account is rather long and quite detailed. Hence, in the interest of brevity, I will give here a summary of the overall experiment, with the hope that enough detail is made available for the reader to be persuaded that the design history of the Britannia Bridge does, in fact, constitute a rather strong corroboration of the hypothesis law. The interested (or skeptical) reader can, of course, always appeal to the larger account for more details!

In the summer of 1844 Parliament passed a bill authorizing the Chester and Holyhead Railway to build a railway bridge across the Menai Strait in Wales. Robert Stephenson (1803–1859), son of the railway pioneer George Stephenson, and himself already distinguished for his contributions to railway locomotive technology and civil engineering, was appointed chief engineer for the project. Almost thirty years earlier, in 1826, Thomas Telford (1757–1834), one of Stephenson's eminent predecessors in British engineering, had built a road suspension bridge across the Menai Straits, and at that time, several points of crossing had been studied. Now, a parliamentary committee used the Telford studies in its hearings on the location and nature of the proposed railway bridge. The committee decided to adopt the site near a rock—known as the Britannia Rock—situated roughly in the center of the channel. Hence the name of the bridge. The width of the Menai Strait at this spot, at high tide, was about 1,100 feet.

The records of the progress of this enterprise, as documented by Clark, Stephenson, and Fairbairn, reveal that the course of reasoning engendered by this project was constituted of four significant and temporally distinct stages:

1. July 1844–May 1845: Formation and development of the initial idea of a tubular bridge.
2. July–August 1845: Performance of a series of experiments with small, laboratory-scale tubes.
3. September–October 1845: Performance of a second series of experiments with laboratory-scale tubes.
4. July 1846–April 1847: Performance of a series of experiments with large-scale-model tubes.

In the first stage, Stephenson, after initially ruling out the suspension bridge form—the most obvious candidate for long-span bridges in the mid-nineteenth century—because of its insufficient rigidity, and the cast-iron arch bridge because it would impede navigation of the waterway, returned to the idea of the suspension bridge, but with the possibility of a stiffer platform. A contemporary engineer, J. M. Rendel, had already used wooden trusses for the purposes of preventing oscillation in the platform of suspension bridges, a fact that was brought to Stephenson's attention. However, he thought that a stronger trussing system would be needed. After considering a number of alternatives, he decided on combining the use of suspension chains with trussed vertical sides and cross braces on the top and the bottom—effectively producing a bridge deck surrounded on all sides by a trussed framework.[24]

However, Stephenson decided against the use of *wooden* trussing, partly for structural reasons and partly because of its susceptibility to fire. Recalling his prior work on a small bridge in 1841 in which *wrought-iron* plates had been riveted together, he decided to substitute riveted wrought-iron plates for the wooden trussing and cross braces. The form that resulted was that of a rectangular *tube* supported by suspension chains *through* which a train would pass.[25] In effect, Stephenson had framed the following hypothesis:

> H_1 A wrought-iron tubular bridge with suspension chains will satisfy the strength, stability, and navigational requirements demanded of the bridge.

The problem with this quite new idea of a wrought-iron tunnel-like bridge, was that there existed neither theoretical nor experiential knowledge that could be applied to the design of such a structure.[26] Being the inventor of a new bridge form, the responsibility lay with Stephenson to convince himself and others that this idea was viable. That is, he was obligated to demonstrate the correctness of the above hypothesis. Stephenson realized that the hollow tube could be viewed as a *beam* in which the top and bottom surfaces would serve the structural role of a conventional I beam's top and bottom flanges. This notion was attractive, since the method of calculating stresses in cast-iron beams under vertical loads was reasonably well established in the mid-nineteenth century.[27] Stephenson had in effect "mapped" a novel and unfamiliar situation onto a situation with which he was familiar[28]—a classic instance of the use of *analogy* in problem solving and reasoning.[29] Thus, because of the intractability (or direct untestability) of hypothesis H_1, Stephenson was led to replace the latter by a more tractable hypothesis

> H_2 A wrought-iron tubular *beam* with suspension chains will satisfy the strength, stability, and navigational requirements demanded of the bridge.

Initially, Stephenson had conceived a rectangular cross-section for the tube. Later, thinking that this section would be structurally inadequate, both in resisting vertical load and the lateral effect of wind forces,[30] he proposed as alternatives circular and elliptical sections, but without actually discarding the rectangular form altogether. Effectively, Stephenson, as a result of critical thinking about H_2, had been led to frame the following auxiliary hypothesis:

AH_1 Circular and elliptical sections are superior to the rectangular section in resisting distortion and side pressure due to wind forces.

This was the situation when, in April 1845, Stephenson consulted William Fairbairn (1789–1874), a prominent iron worker, engineer, industrialist, and engineering scientist, and one who would exert considerable influence on Victorian men of technology through his prolific writings on technical subjects.[31] Fairbairn's main contribution at this stage was to express general confidence in Stephenson's idea.[32]

In May 1845, a parliamentary committee was presented by Stephenson with a design proposal based on the foregoing deliberations. The bridge would consist of two spans, each 450 feet long and 105 feet high, supported by a pier on the Britannia Rock in the center. The tube itself would be 25 feet high. On June 30, 1845, the Chester and Holyhead Railway Bill received royal assent.

The second and third stages each comprised a series of experiments using small, laboratory-scale tubes. These experiments were conducted by Fairbairn and Eaton Hodgkinson (1789–1861), a former student of the celebrated founder of the atomic theory, John Dalton, and an engineering scientist already distinguished for his contributions to engineering mechanics and strength of materials.[33] The reason for carrying out these experiments was that, although there existed, as already noted, a theory that enabled stresses to be calculated in beams of different cross sections under vertical loads, the theory did not allow one to predict the ultimate strength (or breaking load) of a beam of some specified material. Moreover, whatever experimental data there was pertained to cast and not wrought iron.[34]

The first series of experiments, conducted in July and August 1845, involved tubes, 15–30 feet in length, 12–24 inches in diameter, of circular, elliptical, and rectangular sections. Clearly these experiments were intended to test *inter alia* the auxiliary hypothesis AH_1. In fact, Fairbairn began his experiments with circular and elliptical sections.

The tests yielded wholly unexpected results—tubes of all three sectional shapes failed by *buckling* of the upper sides of the tubes, a mode of failure quite unanticipated.[35] In addition, the circular and elliptical tubes were distorted. In effect, the bias toward these two shapes had to be relaxed, and the rectangular section then became the preferred shape. Thus, as a consequence of these experiments, hypothesis AH_1 was replaced by the following auxiliary hypotheses:

AH_2 The rectangular tube section is superior to the elliptical and circular sections in resisting distortion.

AH_3 A wrought-iron tube with rectangular section will satisfy the strength and stability requirements of the tubular bridge.

The buckling problem, however, still remained, even in the case of the rectangular section. It is, perhaps, necessary for the reader to further appreciate the significance of the issue. In the middle of the nineteenth century, the main experience of engineers had been with "I beams"—that is, beams with cross section in the shape of an I; and, according to this experience, failure of the top flange of such beams would occur by crushing rather than by buckling.[36] Having formerly taken the view that the tubular beam was structurally similar to the I beam, Stephenson had discovered, by way of the experiments, that the former behaved somewhat differently from the latter!

Despite the buckling problem, neither Stephenson nor Fairbairn rejected the rectangular shape. Nor, it turns out, did they altogether and immediately abandon the elliptical section, despite its failure in the course of the first series of experiments. These facts bear interesting and unexpected testimony to what some critics of Karl Popper's theory of how science progresses have remarked: that contrary to Popper's claim, scientists—and, in the present case, technologists—do not discard or reject a theory or hypothesis as soon as it is falsified, or just because there is some evidence against it.[37]

The second series of experiments (the third stage of the process as a whole), conducted by Fairbairn between August and October 1845, clearly addresses the auxiliary hypothesis AH_3. As I have shown elsewhere, the reasoning behind the details of these experiments was quite involved.[38] Several additional auxiliary hypotheses were also generated. But, in essence, they entailed the construction and testing of a variety of modified versions of the rectangular and elliptical sections, the modifications involving the addition of material to the top surface of the tube in order to strengthen it. The outcomes of these experiments were, first, that the elliptical section, even after modification, was distorted at a much lower load than the rectangular one—and, hence, was rejected entirely thereafter; and second, that a rectangular section with a *multicellular* top (Figure 6-4) not only did not buckle under load—thereby solving the buckling problem—but also yielded the greatest strength.[39]

In effect, largely as a result of direct physical experiments, hypothesis H_2 had been replaced by a stronger hypothesis

H_3 A tubular wrought-iron bridge with a cellular-topped rectangular section will satisfy the strength, stability, and navigational requirements demanded of the bridge.

Comparing H_3 with its predecessor, the reader will note that mention of the suspension chains has been dropped along the way. In fact, the matter was not quite so simple. While Fairbairn claimed to have been quite sanguine early in his association with this project that the chains were redundant, Stephenson was less assured of this even at the end of the laboratory experiments, according to Fairbairn.[40] And while it is not entirely clear from the records when precisely Stephenson abandoned the idea, it appears, according to Edwin Clark's account, that by the middle of July 1846, the use of suspension chains had been fairly well discarded.[41] Thus, the hypothesis H_3 may be said to represent the general state of affairs sometime between February 1846, when Stephenson submitted his report to the directors of the Chester and

Fig. 6-4. *A rectangular section of the experimental tube used in the course of designing the Britannia Bridge.* The tube was built and tested by Sir William Fairbairn in 1845. The top of the section consists of several hollow cells.

Holyhead Railway, and July 1846, when the fourth and final stage of the design process commenced.

The laboratory-scale experiments had established the overall form of the tubular bridge—what engineers now call conceptual design. There still remained the detailed design to be performed—in particular, the determination of the proportion of the cross-sectional areas of the top and bottom flanges so that they would be equally strong. The purpose of the fourth and final stage was to help establish this objective by way of experiments on a *scale model* of the tube, one-sixth the actual size.[42] The first of these experiments was conducted in July 1846, and the sixth and final one was completed in April 1847. The experiments proceeded with an initial distribution of material in the top and bottom flanges in the ratio of three to one, measuring the breaking load and observing the mode of failure, modifying the scale tube accordingly, to change the distribution of material on the top and bottom, testing again, and so on. With each successive experiment, the ratio of the areas of the top and bottom flanges was progressively reduced and the breaking load was found to progressively increase.

In the last two experiments, the ratio of the two sections had converged to almost one to one. The breaking load had increased from 35 tons approximately in the first experiment to about 86 tons in the last.[43]

This phase of the overall process is a rather good example of what Walter Vincenti called *normal design*.[44] The engineer already knows at the onset the general form of the desired artifact and its general behavioral characteristics; the design issue is to arrive at a specific instance of this known artifactual form so as to meet a specific set of requirements. Such design activity is, very frequently, complex in its own right, but is unlikely to create original form. In normal design, the engineer works well within a paradigm. The parallel with Thomas Kuhn's notion of *normal science* is obvious.[45] The process of designing the tube by way of the scale-tube experiments also appears to be a textbook example of a *search* through a "space" of feasible tubular forms in the sense that this word is used in artificial intelligence. I shall dwell on the matter of search in the next chapter.

As I have already pointed out, the train of reasoning underlying this stage of the process was quite complex and entailed the development of several auxiliary hypotheses which have not been stated here. Details of these have been described in the original paper on the topic.[46] In the course of the experiments some of these hypotheses were corroborated, others refuted and abandoned. But by the end of the third experiment (performed in late August 1846), Fairbairn was quite confident as to the viability of the cellular-topped tubular form. Thus, this experiment provided qualitative support for hypothesis H_3. The fifth experiment, performed in December 1846, and designed to test the lateral strength of the tube—that is, its resistance to wind forces on the vertical sides—provided further evidence corroborating H_3. However, partly as a result of testing and refuting the auxiliary hypotheses, and partly as a consequence of the entire set of model-tube experiments, Stephenson and Fairbairn agreed on a modification to the design, wherein, in addition to the multicellular top flange, the bottom flange was also constructed in multiple cells (Figure 6-5).[47] In other words, a new and final hypothesis emerged, replacing H_3:

Fig. 6-5. *The final sectional form of the Britannia Bridge tube.* Both the top and bottom parts consist of multiple hollow cells.

H_4 A rectangular-sectioned tubular wrought-iron bridge with a cellular top and a cellular bottom (Figure 6-5) will satisfy the strength, stability and navigational requirements demanded of the bridge.

To sum up, the process of designing the Britannia Bridge is found to exhibit the following set of characteristics:

1. It is composed of a single chain of four main hypotheses, H_1, H_2, H_3 and H_4, that were ordered in time. Each hypothesis is associated with a specific form for the bridge: H_1 is associated with a suspended wrought-iron tubular form, while H_2 pertains to a suspended wrought-iron tubular beam bridge; H_3 is associated with a multicellular-topped wrought-iron tubular form; and, finally, H_4 refers to a wrought-iron tubular beam with multiple cells at the top and bottom.
2. Each successive hypothesis in the chain was derived from its predecessor. H_2 was the outcome of critical thinking about the bridge form associated with H_1; H_3 resulted from laboratory experiments intended to serve as qualitative tests for H_2; and H_4 replaced H_3 as a result of the model-tube experiments that were intended as qualitative tests for H_3. The various auxiliary hypotheses generated in the course of the process also played roles in the generation, testing, and modifications of the main hypotheses.

The two corroborating conditions for the hypothesis law appear to have been well met in the case of the Britannia Bridge, as they were in the case of the high-temperature superalloys and the convex-hull algorithm discussed earlier.

7

The Process of Ideation

If the hypothesis law is correct, the construction of mature artifactual form proceeds in a conjectural manner: designs and their associated hypotheses are built, tested, and modified over time. This view of design is reminiscent, in an obvious way, of Karl Popper's celebrated claim that scientific discoveries are effected through cycles of conjectures and refutations.[1] Thus, if the hypothesis law is accepted as a valid law of technological reasoning and one also accepts the essence of Popper's thesis, there would then appear to be a strong similarity between reasoning in science and in technology.

Of more immediate concern is a question that the hypothesis law does not adequately answer. The reader will recall from Chapter 6 that the law was extracted from a knowledge-level model of the design process. And, as the examples in that chapter suggested, the framing, testing, and subsequent modification of hypotheses are obviously dependent on two of the main ingredients of the knowledge level: goals and knowledge. In fact, the reader may realize that my arguments, intended to demonstrate how these cases corroborate the law, appeal entirely to goals and knowledge. In the case of the superalloy problem, for instance, beginning with a *goal* to design an alloy that would have high resistance to creep deformation and hot gaseous corrosion, the Mond metallurgists resorted to their *knowledge* of (i) a possible connection between creep resistance and age hardening, and (ii) evidence that certain nickel-chromium alloys exhibit both properties, to frame the first of the main hypotheses:

> 80/20 nickel-chromium and/or nickel-chromium-iron alloys suitably age hardened would meet the goal.

However, nothing was said in this account of the knowledge-level *actions* that were performed to arrive at this hypothesis—that is, about the nature of the attendant knowledge-level process itself. Similarly, when tests revealed that the creep strengths of different batches of the 80/20 alloy were inconsistent, the metallurgists drew upon their knowledge that such differences could be due to variations in the titanium and carbon contents of the batches. The alloy was modified accordingly, resulting in the first of the superalloys, Nimonic 75. Yet, again, nothing was said of the nature of the process or mechanism that might have been enacted in the minds of the metallurgists that may have led them to this alloy form. We can make similar observations in the case of the Britannia Bridge and the convex-hull algorithm.

So the question is: Can we hope to construct or postulate a knowledge-level process, composed of actions of the kind described in Chapter 4 that would plausibly explain the genesis of a particular technological idea, an artifactual form, or a hypothesis about an artifact?

For want of a better term, let me call such momentous cognitive events *ideations*. Generally speaking, an invention, discovery, literary product, or work of art can entail *many* ideations; on the other hand, there may be only one that is properly significant, producing *the* central idea. The rest of the process of creation may serve only to refine or elaborate the crucial idea. The question, then, is: *Can we explain technological ideation in terms of goals, knowledge, and actions?*

The main message of this chapter is that it *is* possible to explain ideation in terms of knowledge-level processes—at least in the domain of technology and in the natural sciences. But my supporting arguments for this thesis will appear quite late in the chapter—after we have considered what others have had to say about the nature of ideation.

The Classical View: Ineffability

Most writers on creativity have traditionally treated the question of ideation—the act of creation itself—with extreme reverence. Their common opinion, which I shall call the *classical* view, sees ideation as a kind of thinking that is fundamentally different from ordinary, garden-variety thinking. It places great emphasis on the role of the unconscious, on an inexplicable thing called "intuition," on sudden moments of illumination. The classical view supports and nurtures what the psychologist Robert Weisberg wryly dubbed the "myth of genius."[2] And, it has been lent prestige and authority by the fact that many of the most creative men and women who have ever lived have themselves expressed a kind of ineffable wonder as to how they themselves discovered, wrote, painted, composed, and invented.[3]

The Wallas Model

Among the most influential and widely cited explorations of the act of creation within the classical perspective is the mathematician Jacques Hadamard's account of the psychology of mathematical invention.[4] In this work, Hadamard pursues and applies in some detail a line of thinking with a distinguished pedigree—for it seems to have originated in a late-nineteenth-century lecture by the German scientist Hermann von Helmholtz,[5] and further discussed in a famous article written in 1908 by the French savant Henri Poincaré.[6] It was then developed somewhat more systematically by Graham Wallas in his 1926 book, *The Art of Thought*.[7]

According to Wallas (and, later, Hadamard), the creative act—ideation—is composed, broadly, of four steps.

1. *The preparation stage*: The problem is consciously studied, investigated, and pondered from all conceivable points of view.
2. *The incubation stage*: The problem moves from the realm of conscious thought into the unconscious. That is, the person does not consciously think of the problem; rather, an involuntary or unconscious mental process may take place in the course of this stage.
3. *The illumination stage*: Still through involuntary or unconscious mechanisms, the solution to the problem is obtained.
4. *The verification stage*: Conscious thought is resumed for the purpose of revising or developing or refining the solution.

This "model" is more or less in sympathy with the classical view, in the sense that its significant elements are the incubation and illumination stages, both of which are said to operate at the level of the unconscious.

Among the earliest attempts to test Wallas's model were Catherine Patrick's experiments in the 1930s with poets and artists.[8] Patrick used the technique of verbal protocol: That is, she asked her subjects to describe—vocalize—their thoughts as they went about their given tasks. In the case of the poets, the experimental situation consisted of the subjects being shown a picture of a landscape and being asked to write a poem about it. In the case of the artists, they were presented with a poem (a selection from John Milton's "L'Allegro") and were asked to draw a picture about it.

Patrick's conclusion was that, by and large, both poets and artists exhibited all four stages of the Wallas model. However, from the answers supplied by her subjects, Patrick also found that incubation did not necessarily entail an uninterrupted phase during which a subject did not think consciously of the problem at all. Several of the artists reported of an idea that kept "coming back," while they were occupied with other things. More generally, Patrick also noted that the strict linear order of the four stages as stipulated by the Wallas model did not necessarily hold in the case of her subjects. The steps would overlap: incubation often occurred during preparation; and verification (or revision) would begin during the period of illumination.

Given the centrality of the "cognitive unconscious" in the classical view—as manifested, for example, in the introspective reports recorded by such luminaries as Helmholtz, Poincaré, Mozart, and Coleridge,[9] and as explicitly embedded in the Wallas model—Patrick's findings must be regarded with some attention. They seem to undermine the notion that acts of creation or ideation occur during the stages of incubation and subsequent illumination, both of which are postulated as occurring at the unconscious level. In a recent assessment of the evidence concerning the Wallas model in which he reviewed the significance of the unconscious in the creative process, the psychologist Robert Weisberg pointed out that introspective reports or "self-reports" must be treated with some measure of skepticism.[10] For instance, the authenticity of Mozart's famous letter to a friend in which he describes how his musical ideas are born[11] has been questioned by musicologists; and Coleridge's still more celebrated account of the circumstances attending his composition of the fragmentary poem "Kubla Khan" was, according to John Livingstone Lowes, a fabrication by the poet.[12]

Furthermore, after reviewing Patrick's findings and more recent experimental studies, Weisberg concluded that there is little evidence in support of the notion of unconscious incubation as a *distinct and critical* stage of the creative process.[13] Later, I shall return to the issue of the unconscious as a participant in ideation.

There are also other problems with the Wallas model. Consider, for instance, the verification stage, during which, according to Wallas, the validity of the idea or solution obtained during illumination is tested and further refined to a more precise form. In incorporating verification as a distinct stage in his model, Wallas was clearly influenced by Poincaré's celebrated description of the phases in two of his own mathematical discoveries. For Poincaré, the unconscious illumination stage produces ideas which then has to be verified by disciplined, attentive—and consequently conscious—calculation.[14]

However, there is no guarantee that the idea, the illumination produced in the unconscious, according to Poincaré, Wallas, and Hadamard, is in any way correct—that all it subsequently engenders is elaboration or refinement. Bounded rationality puts paid to such assurance. If, during the so-called verification phase one discovers the original idea to be wrong, then one must return to the drawing board—in the case of technological creation, literally so! This is precisely what the hypothesis law asserts in the case of technology, and there is every reason to believe that it is likewise in other kinds of creative acts. Of writing fiction, the novelist and teacher of creative writing John Gardner has written:

> At some point, perhaps when he's finished his first draft, the writer begins to work in another way. He begins to brood over what he's written, reading it over and over, patiently, endlessly, letting his mind wander. . . . Reading in this strange way lines he has known by heart for weeks, he discovers oddities his unconscious has sent up to him, perhaps curious accidental repetitions of imagery. . . . Why? he wonders. . . . The writer assumes that the accidents in his writing may have significance. He tries various possibilities. . . . He makes tiny alterations. . . . Slowly, painstakingly, with the patience that separates a Beethoven from men of equal genius but less divine stubbornness, the great writer builds the large, rock-firm thought that is his fiction.[15]

In other words, writing fiction, according to Gardner, is to a great and critical extent an act of rewriting. An extreme and dramatic example of this from the realm of poetry, is the history of William Wordsworth's long autobiographical poem "The Prelude." This poem, unpublished during Wordsworth's life, exists in manuscript in some four different versions composed in 1798, 1799, 1808, and 1850, the year Wordsworth died. "The Prelude" was, thus, composed over a period *in excess of half a century!*[16] Furthermore, as Linda Jeffrey has illustrated, the transitions between these versions were highly convoluted, entailing not only expansion of an idea or feeling, but omissions of lines or sections from one version to the next, introduction of new ideas and reintroduction of lines or ideas discarded in a previous version.[17]

It is, thus, hard to sustain the belief that verification, in Wallas's and Hadamard's sense, is to be distinguished as a conscious phase that can be held apart from the other phases they write about. Rather, to echo Catherine Patrick's empirical findings,[18] one can at best conclude that preparation, incubation, illumination, and verification *are not stages at all* in creative thinking but that these phenomena are *ingredients* in the process of ideation and that they are inextricably intertwined. If this conclusion is correct, we have to admit that the cumulative wisdom of Helmholtz, Poincaré, Wallas, and Hadamard provides no clue as to the nature of the *process* of ideation. We need to explore other avenues.

Bisociation: The Combination of Unrelated Ideas

One view that appears to be very widely held by thinkers on creativity is that the essential ingredient in ideation is the effective combination of possibly disparate or unconnected ideas—"elements drawn from domains which are far apart," in Poincaré's words.[19] Arthur Koestler featured this notion in his book *The Act of Creation* under the name of *bisociation*.[20]

Like Poincaré and Hadamard, Koestler associated creativity with unconscious thought—more precisely, with the unconscious combining of ideas. But bisociation is not merely the combination of ideas but of ideas from different "planes," from "frames of reference" or—the term Koestler seemed to favor most—"conceptual matrices." By matrix, Koestler meant "any ability, habit or skill, any pattern of ordered behavior governed by a *code* of fixed rules."[21] Bisociation entails the linkage of (at least) two normally incompatible or unrelated matrices.

Whether or not ideation—leading to psychologically or historically original thought products—involves the combining of unrelated ideas is largely an empirical issue. One needs only to examine historical cases to test this hypothesis, and there is indeed sufficient evidence in its support. Restricting ourselves to the realm of technology, for example, I have pointed out in my book *Creativity in Invention and Design* that Maurice Wilkes's invention of microprogramming, which signifies both a technique for designing the control unit of a computer as well as an architecture for control units, entailed the combination of two entirely unrelated concepts—namely, the concept of programming on the one hand and a particular kind of circuit structure called the diode matrix on the other.[22]

In the case of Alexander Graham Bell's development of the telephone, Bernard Carlson and Michael Gorman have documented, after examining Bell's notebooks, drawings, and the relevant artifacts, how his first telephone, the so-called gallows telephone of 1875, was the complex, emergent outcome of several artifacts and ideas that were essentially unrelated to one another: Hermann von Helmholtz's experimental apparatus (using tuning forks) for the study of the production of vowel sounds, his own experience with the "phonoautograph," a device for teaching the deaf to recognize speech patterns—Bell was a teacher of the deaf and a professor of elocution at Boston University—and his knowledge of the structure and functioning of the human ear.[23]

And, as perhaps the most striking instance of bisociation at work, the aeronautical historian Tom Crouch has explained in some detail how Wilbur and Orville Wright resolved the problem of aircraft stability and control by combining three very distinct ideas. In Crouch's own words: "The Wrights had taken a set of graphic images—a bicycle speeding around a corner, a bird soaring through the air, a cardboard box twisted in the hands—turned them into thought problems and reassembled the lessons learned into a mechanical system for controlling an airplane in the roll axis."[24]

Yet, despite what seems to be the essential correctness of Koestler's concept of bisociation, he leaves unexplained the all-important matter of the *mechanism* of bisociation. How can different ideas come together, combine, and produce an idea or concept that is original?

The Darwinian Model: The Variation-Selection Theory

The most famous, studied, and discussed process whereby novelty comes into being is evolution by natural selection, in which the underlying mechanism that allows natural selection to operate is genetic variation. Ever since Charles Darwin and Alfred Russell Wallace (more or less independently) proposed the theory of evolution in the 1850s, and even during the period when the genetic basis of natural selection was unknown, thinkers have sought to apply Darwinian ideas (unfortunately for Wallace, this is the adjective that has prevailed) to domains other than the origin of species. And, given that the principle of natural selection pertains to the *creation* of species, it should not surprise us that those who ponder the nature of cognitive creativity have also strove to apply Darwinian ideas to the resolution of their particular problem.

In 1960 the psychologist Donald Campbell proposed a Darwinian theory of the generation and growth of knowledge.[25] Through Campbell's later writings and those of other writers this approach to the "problem of knowledge" has come to be known as *evolutionary epistemology*.[26] In recent times, Campbell's theory or model has served as a basis for the construction of how both scientific discoveries are made and new technological knowledge is generated— and, here, the latter includes both theoretical knowledge of technological principles and the knowledge embedded tacitly in artifactual forms. In 1988, Dean Keith Simonton, a psychologist, proposed a Darwinian model of the psychology of scientific discovery.[27] Two years later, Walter Vincenti, an aeronautical engineer and historian—whom we have already encountered in connection with the Britannia Bridge—published a similar model to explain the generation of engineering knowledge.[28] Given that Vincenti does not appear to have been aware of Simonton's work (at least, there are no references in Vincenti's book to the latter's publications), we cannot fail to be impressed by this instance of a "multiple discovery" involving scientists from two very different backgrounds.

Let us then examine what light the Darwinian perspective sheds on creativity—in particular, technological creativity.

Campbell's argument may be summarized in the following terms.

First, the production of genuinely new knowledge or ideas—thought products that are psychologically or historically original, as characterized in Chapter 5—demands the generation of *variations* that are *blind* in the sense that the consequences of these variations go beyond what can be foreseen or anticipated.

Second, the variations are then subjected to a *selection* process that prunes out all variations but those that demonstrate a *fit* with the problem at hand. Such a selection process in the context of scientific discovery demands one or more criteria whereby the alternative solutions can be judged to have or not have solved the problem. Those variations that fail to meet the criteria adequately are rejected.

Third, there must be a mechanism for the *retention* (and propagation) of the selected variation. For, without retention, a selected or successful variation can hardly make a permanent contribution to the acquisition and growth of knowledge.

For those familiar with the theory of organic evolution by natural selection, it is clear that Campbell's model is drawn directly from the Darwinian paradigm. For our present purposes, the last component, retention, is not important (at least in this chapter) since my concern for the moment is with the process of ideation itself. Thus, henceforth, ignoring the issue of retention and following Vincenti, I shall call this the *variation-selection theory*. The issue then is: Does this theory provide a plausible basis for explaining how original ideas in science or technology are born? And if so, what are the mechanisms for variation generation and selection?

Simonton's Model

Simonton's version of the variation-selection theory—which he calls the *chance-configuration* theory—attempts to outline, in some detail, the cognitive mechanisms underlying variation and selection.[29] His account can be concisely paraphrased as follows:

1. The creative process involves operations on *mental elements*—the "fundamental [psychological] units that can be manipulated in some manner,"[30] such as sensations, emotions, ideas, concepts, and so on. In the case of scientific creativity, the primary mental elements are "cognitions of some kind, such as facts, principles, relations, rules, laws, formulae, and images."[31] These mental elements may be held in the unconscious as well as the conscious state, and may be invoked voluntarily or involuntarily.

2. The basic generating mechanism in scientific creativity is the *chance permutations* of those mental elements—"permutation" in the mathematical sense that the *order* in which the elements enter into combina-

tion matters; and "chance" (Simonton's word for Campbell's "blind") in the sense that which mental elements combine is unpredictable—but not necessitating total randomness. In Simonton's words, "We do not need to argue that all permutations of a specific set of elements are equiprobable. . . . We must merely insist that a large number of potential permutations exist, all with comparably low but nonzero probabilities."[32]

3. Not all chance permutations of mental elements are equally stable. Those permutations that lack sufficient coherence to form stable bonds may simply be referred to as *aggregates*. Those that do cohere and form stable structures constitute what Simonton termed *configurations*. And while permutations may be subject to chance, the formation of a configuration is determined by the properties or characteristics of the constituent elements and how these properties of the different elements are mutually compatible or show mutual affinities. Simonton, here, draws on an analogy with chemical bonding where the valence property of the atoms determines the formation of specific molecules.

4. Configurations are of two kinds. *A posteriori* configurations between mental elements are formed when the corresponding real-world events that they signify or represent have a high probability of co-occurrence. Such configurations arise from experience. If, for example, whenever nickel is used as an alloy base, we observe that the alloy resists the physical phenomenon of creep (see Chapter 6), the mental elements corresponding to "nickel-based alloy" and "creep," respectively, will form an a posteriori configuration.[33] In contrast, *a priori* configurations are derived from conventions of the kind prescribed in mathematics, logic, and even law. Simonton's examples are somewhat vague in this regard, but, if I have interpreted him correctly, rules of algebra such as the commutative and distributive laws—

$$A \times B = B \times A$$
$$A \times (B + C) = (A \times B) + (A \times C)$$

—are instances of a priori configurations where the mental elements entering into combination are the expressions on the two sides of the equals sign.

5. Configuration formation, then, is the means by which variations—clusters of mental elements—are selected and, to some extent, retained. In Simonton's theory, however, retention demands a further state of what he calls *communication configuration*, where the configuration previously formed in the agent's mind is structured or reconfigured in a way suitable for communicating to others. In other words, while a configuration, when formed, by itself constitutes personal knowledge, its retention and transmission demand conversion to public knowledge. This may entail a reconfiguration, as it were, for the purpose of public expression.

In the light of what I have said earlier about psychological and historical originality, it is clear that while (internal) configuration formation may suffice for the agent to assess whether or not the thought product is psychologically original, assessment of historical originality certainly demands something corresponding to Simonton's notion of communication configuration.

As evidence in support of his theory, Simonton relies in part on the impressionistic and introspective reports of the kind Brewster Ghiselin had assembled in *The Creative Process*.[34] Simonton quotes extensively from the writings of Poincaré, William James, Einstein, and Hadamard. He then goes on to describe his own and other psychologists' findings on personality, cognitive, and motivational traits of creative scientists and suggests how such traits are consistent with the chance-configuration model. Later in his book, Simonton deals with other external (that is, historical and sociological) features of the "scientific genius" as, for example, their generally high level of productivity, family and cultural backgrounds, intellectual influences, education, academic success, and the scientific Zeitgeist, and explains how these characteristics, as observed by psychologists, sociologists, and historians, are explicable in terms of the chance-configuration theory.

Simonton has also examined the phenomenon of multiple discovery—when two or more scientists, independent of and unknown to one another, discover or invent something almost simultaneously. Simonton argues, on probabilistic grounds, that the phenomenon of multiple discovery is indeed explainable in terms of the variation-selection model. Unfortunately, he does not offer any empirical evidence in the form of actual historical cases of scientific discovery in support of his theory.

Vincenti's Model

Like Simonton, Walter Vincenti begins with Campbell's thesis; unlike Simonton, however, his main contribution is to demonstrate, using several case studies from the annals of aeronautics and aviation, how the variation-selection theory serves to explain the generation of technological knowledge. Furthermore, Vincenti is less concerned with what we might regard as the highest acts of technological creativity—what I have earlier called PO-creative or HO-creative acts. Rather, Vincenti's account is primarily concerned with what the historian of technology Edward Constant, following Thomas Kuhn's lead, called *normal technology*—that is, "the improvement of the accepted tradition or its applications under new or more stringent conditions."[35] Vincenti calls the activity of design within the realm of normal technology *normal design*.[36]

In Vincenti's view, the prime sources of variations (of designs, technological ideas, or artifactual forms) include at least the following: prior experience or knowledge; the incorporation of possibly arbitrary novel features into a given design; and the "mental winnowing of the conceived variations to pick out those most likely to work."[37] With regard to the second of these, Vincenti does not amplify how the novel features are generated and, to this extent, seems to beg the very question of how variation arises. But more curious is the third aspect, for here he seems to include in his catalog of variation mechanisms what one would usually regard as a selection mechanism.

As regards the means of selection, Vincenti identifies two: the use of experiments or simulation; and analysis and calculation.

In other words, Vincenti's version of the variation-selection model is not confined to ideation alone. In fact, it seems to be only marginally concerned with ideation. For example, one of Vincenti's case studies concerns the development and evolution of an airfoil (that is, aircraft wing) design that took place within one particular company, the Consolidated Aircraft Corporation of San Diego, and involved a particular engineer, David R. Davis.[38] The period of concern was from approximately 1934, when a patent for a family of airfoil shapes was issued to Davis, to about 1942. The episode itself involved not only extensive wind-tunnel testing of the Davis wing but also its incorporation into three Consolidated aircraft, the Model 31 flying boat completed in May 1939, the B-24 bomber completed in December 1939, and the B-32 strategic bomber commissioned in 1942.

In sum, Vincenti's variation-selection theory applies to the same cognitive level as does the hypothesis law of maturation. It would seem that the primary difference between variation-selection and the hypothesis law is that the former, being Darwinian in scope and spirit, implies that *several* variations are generated, in contrast to one hypothesis or a very small number of alternative hypotheses formulated in any given cycle, as is implied by the latter.

We have already seen examples of such hypotheses generated in the case studies discussed in Chapter 6. Later in this chapter I shall reconsider these same problems in the context of the variation-selection theory. But to illustrate this model in action taking one of his own examples, Vincenti writes of the Davis wing design:

> The variation-selection process in design appears clearly in the work at Consolidated Aircraft. In seeking the wing to use on their long-range airplanes, the company's engineers examined overtly a large number of variations (and doubtless imagined still more). Selection depended on . . . trial by a combination of analytical study and wind-tunnel test. The variations had a good deal of blindness, and the wind-tunnel methods were far from sure; the combined level of uncertainty was high. Direct trial with the B-24 was completely unsure in its ability to separate out the performance of the Davis airfoil; selection in the environment was in the end inconclusive in this respect.[39]

The variation-selection theory, as specified by Vincenti, does not tell us much about how a particular variation—a particular idea—actually comes about just as the hypothesis law is silent in this regard. Neither provides insight into ideation—which is what the concern of this chapter is. And Simonton's chance-configuration model, though more fine-structured and applicable to the process of ideation, falls short of explaining how configurations of mental elements may come about.

How Valid Is the Darwinian Model?

At the very core of the theory of evolution by natural selection is the idea that for evolution to take place there must be *a large number of genetic variations* on which natural selection can work. In biological terms this means that organisms must be "superfecund" in their reproductive capabilities; that is, the population of organisms will, in general, be much larger than what the availability of food and space will allow. It is the large genetical variations in the offsprings of organisms that make it possible for selection to occur. The significance of superfecundity in Darwin's actual thinking-out of his theory has been beautifully recorded by Howard Gruber in his study of Darwin.[40] Its centrality in the mechanism of natural selection is now a fact of modern biological thought.[41]

If the Darwinian model is to hold in the case of ideation—if, that is, one is making a claim that the generation of new knowledge, ideas, artifactual forms, or theories is constituted of a Darwinian process—the following questions arise: Is there, in fact, any evidence that the cognitive agent does indeed generate a large number of variations à la Darwin? And, in the context of technology, how can one reconcile the idea of variation generation—that is, the production of a population of alternative artifactual forms or designs—with that of a single hypothesis (or, at best, a very small number of alternative hypotheses), which, the hypothesis law stipulates, is formed at any given time or in any given cycle of the design process?

Such questions are far from easy to resolve. Let us, however, revisit the episodes we examined in the preceding chapter and also take a look at other episodes not hitherto studied to see whether they shed some light on these questions.

Superalloys, Revisited

In the case of the invention of the creep-resistant superalloy, the very first hypothesis that seemed to have been postulated was:

H_1 80/20 nickel-chromium and/or nickel-chromium-iron alloys suitably age-hardened could meet goal G_1.

where the goal G_1 was to produce an alloy that, better than other alloys, satisfies the requirement:

R_1 The alloy must be resistant to creep deformation and rupture at very high temperatures (in the range of 750–1,000°C); it must be resistant to hot gaseous corrosion and to accidental shocks; it must be capable of being formed.

According to Cahn's account,[42] the metallurgists at the Mond Nickel Company posed with R_1 arrived at the *idea* embodied in H_1 because (a) the 80/20 nickel-chromium alloys were already known for their properties of oxidation (that is, corrosion) resistance; (b) there was some prior evidence that nickel-chromium alloys were creep resistant; and (c) there was some evidence of a connection between age hardening and creep resistance. There is no indication, at least in Cahn's paper, that a large number of alternative candidate alloys—a large number of aggregates or partial configurations, to use Simonton's terms—were generated, even transitorily. This in spite of the fact that both age hardening and creep were poorly understood in the late 1930s, and, thus, the metallurgists would have been quite justified in clutching any promising straws.[43]

Rather, the investigators appeared to use the given goal and the knowledge they possessed to select two classes of alloys as possible candidates. Artificial intelligence researchers are familiar with this phenomenon: They would refer to it as using heuristic and other kinds of knowledge to *prune the search space.* That is, given the potentially unbounded "space" of alloys—this corresponding to the space of potential aggregates in Simonton's terms—the metallurgists did not actually search or generate any significant region of the space. Their knowledge and the goal enabled them to home in on a minute part of the space, that is, one containing the nickel-chromium class of alloys. Fecundity in the variations generated seems hardly in evidence here.

If we reflect on the transition from hypothesis H_1 to the next hypothesis—

H_2 Nimonic 75: an alloy such that $G1$ is achieved

—there *is* some evidence of the variation-selection process at work here. Several commercial batches of the 80/20 nickel-chromium alloy (the variations) were tested (subject to selection) for creep strength. However, even here, the process was not that straightforward. It was not a matter of simply selecting from the different batches of this alloy (the batches differing in the amount of impurities) one that had the highest creep resistance. The tests were not consistent, this being attributed to differences in the titanium and carbon contents of the batches. While Cahn's account is not clear on this point, it may be the case that the alloy that emerged, Nimonic 75, containing small amounts of titanium (0.4%) and carbon (0.1%), may in fact have resulted from generating variations based on varying quantities of these elements and testing the resulting alloy for creep strength. If this is the case then clearly some kind of variation-selection process occurred *because there was no knowledge to control or limit the amount of search.*

We also see how the hypothesis law can relate to variation-selection without contradictions: *the hypothesis* H_2 *emerged as a result or outcome of a variation-selection process.*

The remaining steps in the development of the first superalloys, as I described in Chapter 6 using Cahn's historical account as source, are not sufficiently detailed for us to draw any definite conclusions as to the presence or absence of the variation-selection principle. Suffice to say that there is *some* definite evidence of its presence in later steps. For example, as I have summarized the process, there was one point when the newly developed alloy—more precisely, a family of alloys in a compositional range—was subjected to further testing, using X-rays, for its age-hardenability characteristics. Three important conclusions relating age hardening to specific nickel-, aluminum-, and titanium-based phases—these, being the auxiliary hypotheses AH_6, AH_7, and AH_8—were thereby established.

However, and it is important to note this, as far as one can tell from Cahn's description, these hypotheses (or ideas) *were not selected from a repertoire of alternative hypotheses* (or ideas). Rather, they represent *inferences* drawn from the X-ray studies. Precisely what kind of inferences they might be I shall describe later in the chapter. Thus, we may conclude that if we regard these auxiliary hypotheses as stable configurations (in Simonton's language), they were not selected out as the Darwinian theory suggests but were produced by a very different mechanism than the variation-selection process.

The Convex-Hull Algorithm, Revisited

In the case of the design of the convex-hull algorithm, also discussed in Chapter 6, we actually have a record of the overall conscious thought process based on the protocol recorded by Elaine Kant and Allen Newell.[44] In this case, posed with the problem of designing an algorithm to find the convex hull of a given set of points, the protocol yields no evidence of the variation-selection principle at work except at one stage of the process (which I shall discuss below). Rather, the agent—Kant and Newell's experimental subject—appeared to identify very quickly, within a few seconds, a *single* initial idea which in Chapter 6 I referred to as design D_1:

> Select a point and test whether it is on the convex hull or not.

The protocol further reveals that the remainder of the design process follows the hypothesis law in that the basic idea or hypothesis is critically examined and modified, new hypotheses or ideas result, and so the process goes until a satisfactory algorithm is produced. However, at one stage in the process— where the agent attempts to identify a method for establishing a "good" starting point for the algorithm—we see some suggestion of variation exploration. The agent has already identified one point in his test figure (point A in Figure 6-2) as a good starting point but does not know how the algorithm would pick such a point. He decides to choose a second point as an alternative:

> If I start at point B [see Figure 6-2] and I go to A then, either route there, I still have that problem and I want to retreat back to point B. But it turns out that no matter what . . . which direction I go in from point B I'm going to have the same problem. So point B can't possibly be on the convex hull.[45]

Since the alternative fails, the subject abandons these explorations (and the problem itself) for the time being.

Thus, insofar as the conscious trace of Kant and Newell's subject's thinking reveals, the design of the convex-hull algorithm entails, *in one of its stages only,* the barest suggestion of the use of variation-selection. For the rest, the agent's design process appears to entail a more or less single line of thinking, wherein one rough design idea based on the problem itself is first generated then successively tested and refined to another single but more concrete design.

The Darwinian picture *does* emerge with some clarity when we consider the overall experiment involving this particular subject. Kant and Newell have recorded that having produced an algorithm, the subject $S2$ finds that it is not particularly efficient. $S2$ then goes on to produce a second algorithm using an altogether different strategy of algorithm design known in computer science as "divide and conquer." Thus, suppose a historian or philosopher of technology was to examine how, given the problem to design a convex-hull algorithm, this particular subject produced a "final" solution—and this happened to be the divide-and-conquer algorithm; the record would actually show that the agent produced two "variations," and from these the divide-and-conquer version was selected because it was more efficient. The design process viewed on a longer time scale *would* appear to satisfy (a limited form of) the variation-selection process.

In other words, the cognitive act of creation can be meaningfully considered in terms of different *time grains*. This is unsurprising since all processes one is aware of, whether natural or artificial, bear this property. The "act" of invention or design can be taken to mean, on the one hand, the act of producing a single, possibly tentative, possibly incomplete idea—the act of ideation itself. The production of a single hypothesis, say. The time for such ideation may be as small as a few minutes or even less. On the other hand, if by "act" we mean how a "final," mature idea is brought about, then the duration of the act is likely to be considerably longer. In general, we find that the variation-selection process is only one possible mechanism whereby creativity is manifested. It is far from being *the* process, as Simonton's and Vincenti's writings would suggest. However, I surmise that when we consider a longer time scale and take a more macroscopic trace of how an artifactual form or an idea may have been produced, there is greater manifestation of variation-selection at work. The aeronautical cases that Vincenti has presented as examples in support of his theory—and these include the design of the Davis wing previously mentioned, the development of a method of fluid analysis,[46] and the development of retractable airplane landing gear[47]—are all of this breed.

The Britannia Bridge, Revisited

If we consider once more the circumstances leading to the initial idea of the tubular bridge, we see that the two best-known bridge forms of the time, the suspension and cast-iron-arch forms, were both initially considered but then ruled out for reasons I explained in Chapter 6. But Stephenson returned to the idea of a suspension bridge and thought of stiffening the platform; he considered using the trussing method of his contemporary J. M. Rendel but felt that it would not be adequate. By Stephenson's own account, he then considered "a variety of devices" to achieve the stiffening[48]—unfortunately, he did not elaborate on what these were—and eventually decided that the most promising approach would be a combination of the suspension chain and trussing, the latter forming the vertical sides with cross bracing at the top and bottom: the form would then be "a roadway surrounded on all sides by strongly trussed framework."[49]

One can certainly see that the idea of the tubular structure was the result of combining—bisociating—three relatively distinct technical ideas: the suspension chain, the wooden truss, and the use of wrought-iron riveted plates. And clearly there is some suggestion of the variation-selection process at work. Stephenson, according to his account, *did* consider alternative forms; he *did* select from the variations. However, those variations that we are made aware of—those that are recorded, the suspension bridge and the arch forms—are hardly permutations in Simonton's terms; and it seems implausible to think of the final tubular form as being a chance configuration that was selected from the variations. Such a conclusion would simply ignore the fact that there was a *process* involved—a stiffened form of the suspension bridge was first conceived, a "variety of devices" for stiffening the bridge were then considered, from which the idea of wooden trussing of the vertical sides with cross braces at the top and bottom emerged, and this idea was then transformed to and replaced by another form. To explain this process in terms of variation-selection alone is to sweep all the complexity of ideation under the evolutionary carpet.

Ideation as a Knowledge-Level Process

We have seen, so far, several views or models of the ideation process: First, the classical stance, according to which ideation—the act of creation itself—is mysterious, wondrous, ineffable; suffice to say that anyone wishing to make a systematic study of creativity will take a dim view of this stance. There is then the four-stage Wallas model, with its formidable pedigree, which acknowledges the place of both conscious and unconscious states but emphasizes strongly the role of the latter insofar as the act of ideation itself is concerned. Koestler's notion of bisociation ascribes the act of creation to the joining, in some way, of distant and unrelated ideas or "matrices" of knowledge. Finally, there is the variation-selection model, according to which ideation, at least in science and technology, entails the formation of a population of variations of ideas or artifactual forms by chance permutations of knowledge tokens, followed by the selection and retention of one or a few variations.

I hope the reader will agree that, leaving aside the classical view, there is sufficient evidence to draw the following conclusions concerning the other three models: First, they all provide important insights into, and identify some of the ingredients of, the process of ideation. Second, no one of them is complete or sufficient in itself to explain ideation. And, third, no one of them comes to grips with the *mechanics* of ideation: They all leave unanswered the question, *what happens in ideation?*

It is in regard to this question that, I think, the knowledge-level model of cognition presented in Chapter 4 becomes particularly useful. Let me, at this point, propose the following:

Knowledge-Level Hypothesis on Ideation (KLHI): Let ψ be a scientific or techno-logical goal for which π is a solution produced by an agent by way of a cognitive process. Furthermore, let π be novel or original in the psychological or historical sense. Then the process constructed by the agent can be specified by or explained in terms of a knowledge-level process with ψ as input and π as output.

Before we consider how we can test this hypothesis, several remarks are in order.

First, the reader will take note that KLHI is a hypothesis about ideation in the realm of the natural sciences and technology. It is conceivable that the claim made here can be extended to other obvious domains of creativity such as writing and art. However, the scope of the hypothesis, insofar as this discussion is concerned, is the realm of science and (in particular) technology.

Second, KLHI has its origins in the pioneering and celebrated work on problem solving done by Allen Newell, Herbert Simon, and their associates in the late 1950s[50]—work that contributed significantly to the birth of both artificial intelligence (AI) as a field of study and to what is now called the information-processing paradigm for cognitive psychology.[51] In fact, the substance of KLHI is contained in even more emphatic terms in a paper on creative thinking by Newell, Simon, and their collaborator Cliff Shaw,[52] in which they stated that if one could build a computer program that behaved in a manner that would be deemed creative if exhibited by humans, then the principles of such a program would constitute a satisfactory theory of creative thinking.[53]

In the particular form in which I have stated the hypothesis here, that is, in terms of the knowledge level, KLHI was presented by this writer in his book *Creativity in Invention and Design*.[54]

Third, KLHI is *testable* in the following sense: If for a given act of techno-logical or scientific ideation, we are unable to construct a knowledge-level pro-cess that explains it—that is, produces from known initial conditions an out-put representing the idea—or the process so constructed uses certain knowledge tokens that can be shown to have not been part of the agent's knowledge body, then we will have *failed to corroborate* the hypothesis. A single failure may not cause us to immediately reject the hypothesis but it will certainly weaken our belief in it; and if several such failures to corroborate occur then we may in-deed be forced to abandon KLHI. On the other hand, every empirical instance of ideation that *is* explicable in the form of a plausible knowledge-level process will serve to corroborate the hypothesis and thereby strengthen our belief in it. KLHI is, thus, to be squarely regarded as an empirical, testable proposition.

Finally, KLHI is pertinent to a thought product that is original in any one of the four senses described in Chapter 5. In other words, ideation is associ-ated, in this hypothesis, with thinking that may be PN-, HN-, PO- or HO-creative—producing ideas or concepts that range from being simply novel to the originator to ones that are original for the relevant scientific or technological community.

The strongest kind of supporting evidence for KLHI will clearly entail the construction of a plausible knowledge-level process for an act of creation that is unequivocally deemed historically original. For such a process to be constructed, one needs access to a formidable body of information about the agent (or agents) and about the domain of ideation itself, so that one can ascribe a plausible knowledge body for the agent, and can draw upon this knowledge body to construct the process. It may necessitate the use of many different sources of information, including biographical material; historical accounts of the invention or discovery itself; information regarding the state of the public knowledge about the domain at that point of historical time; journals, diaries, and notebooks; personal accounts by the originator of the relevant event; and so on. As the reader may suspect, such detailed knowledge of an agent's background knowledge is not easily gleaned. Fortunately, in very recent years, a few studies from the scientific and technological domains have been conducted along such lines. These include Deepak Kulkarni and Herbert Simon's development of a computer program that reconstructs how the Nobel Prize–winning biochemist Hans Krebs discovered the ornithine cycle for urea synthesis;[55] the series of programs built by Patrick Langley and his associates that "discover" a variety of well-known quantitative and qualitative laws in physics and chemistry;[56] Paul Thagard's computational models that shed light on how theories such as the wave theory of light and major scientific revolutions such as the chemical and geological revolutions might have come about;[57] Bernard Carlson and Michael Gorman's explanation of the stages in the invention of the telephone within a cognitive framework;[58] Robert Weber's identification of "heuristics for invention" whereby the development of the Swiss army knife can be elucidated;[59] and my own account of how the invention, by Maurice Wilkes, of the microprogrammed control unit for digital computers can be explained in terms of a knowledge-level process.[60] Neither Carlson and Gorman nor Weber, however, probe the actual process of their respective inventions at the level of detail required for a test of KLHI.

The Invention of Microprogramming as a Knowledge-Level Process

In Chapter 5, I outlined very briefly the background to Maurice Wilkes's invention of microprogramming in order to illustrate its originality (Figure 5-1). Earlier in this chapter, I also pointed out that, according to the available evidence, the idea of microprogramming appears to have been the fruit of a bisociative process involving two quite unrelated concepts—that of programming and the stored-program computer on the one hand and a circuit form called the diode matrix on the other. That Wilkes's development of microprogramming was an HO-creative process seems indisputable.[61]

In my book *Creativity in Invention and Design*, I conducted a full-length study of the invention of microprogramming as a knowledge-level process in order to test KLHI. My approach was to use the available historical evidence as recorded in the original papers, in subsequent retrospective accounts by Wilkes, in his autobiographical memoir, and through other sources, including personal discussions with Wilkes.[62] Using such documentary records of this historical episode, I was able to construct an explanatory knowledge-level process whereby the invention of microprogramming might have come about.

As already noted previously, the plausibility of such a construction hinges critically on how closely one adheres to the recorded evidence and the believability of the assumptions one makes about the state of the agent's knowledge body. For example, in this particular case study, one could find no evidence that Wilkes explored alternative avenues—that is, that he proceeded in any stage of his activity along the lines of the variation-selection model. Indeed, according to his own statement to this writer on this matter, he could not recall performing any such searchlike procedure; nor is there any record of errors or false trails. Thus, my construction of the knowledge-level process could mirror neither of these facets of creative problem solving.

In this account, I will present only the salient features of the early stages of the overall process that has been constructed and described in great detail in *Creativity in Invention and Design*.[63] My objective here is to give the reader an idea of what a knowledge-level process purporting to model an actual act of technological ideation might look like, and a sense of how and why this particular process serves to corroborate KLHI.

In 1949, Wilkes and his colleagues in what was then called the Mathematical Laboratory in Cambridge had successfully completed the design and construction of the EDSAC. Almost contemporaneously, Wilkes's compatriots at the University of Manchester, led by F. C. Williams and Tom Kilburn, were putting into operation the first of the "Manchester computers." These two machines were the world's first *fully operational stored-program computers*, although a few others were then in various stages of design or construction.[64]

The issue that caught Wilkes's attention at this time, and which marked the beginning of the microprogramming story, was the irregularity and ad hoc nature of the *control circuits*—or as they came to be known, the *control units*—in the few computer designs then in existence. As the term suggests, the control unit is that part of the computer responsible for issuing the proper control signals at the proper times to the rest of the computer during the computer's operation. The irregularity of the control unit contrasted unfavorably with the systematic and regular structure of the other two main components in a computer, namely, the memory unit and the arithmetic unit.

The irregularity and unstructuredness of the control unit became issues for Wilkes because they made control units unduly complex; such complexity detracted seriously from such desirable and practical features as repairability, maintainability and reliability—features that are the hallmarks of "good" engineering practice. These issues were of particular concern to computer designers of the time because of the unreliability of the digital devices then available.

Thus, Wilkes began with an empirical "fact," a generalization founded on his experience with the EDSAC and his knowledge of the Manchester and other computers being developed circa 1949–50:

F_0 Existing control units are difficult to maintain, repair, and design because of their complexity.

The overall process itself consists of a sequence of subprocesses of which the first few are the following:

Problem Recognition

One of the characteristic features of the creative mind is the ability to recognize a problem.[65] This was certainly true in the case of Wilkes: He did not tackle an "open problem" known to the computer community of the time; he *saw* a problem others had not perceived; it originated in the fact just stated and led, first, to a goal

G_0 Develop a control unit form that is easy to design, maintain, and repair.

Thus, this first subprocess is an instance of a *nonrational* knowledge-level process (see Chapter 4) in that there were no identified goals serving as inputs to this process. Furthermore—though this is not intrinsically important to the knowledge-level model of the creative process—it seems reasonable to speculate that this subprocess, causing a transition from a fact F_0 to a purposeful goal G_0, may have occurred at the unconscious level.

Problem Formulation

If problem recognition is one significant marker of the creative persona then so is the ability to formulate the problem in a tractable or solvable way.[66] Goal G_0 as an identified problem is not sufficiently precise, and then are no doubt many ways in which it could be reformulated. In Wilkes's case, the historical evidence suggests a subprocess that began with G_0 and resulted in the goal

G_1 Design a control unit that has a simple, regular, and repetitive structure.

Arriving at the Idea of Using a Diode Matrix

The idea of using a circuit form called the diode matrix appeared to be the first of the significant steps toward the achievement of goal G_1. A diode matrix (Figure 7-1) is basically an array of intersecting "horizontal" and "vertical" wires in which the former receive the input signals to the circuit and the latter transmit the output signals from the circuit. The points of intersection between the orthogonal wires serve as sites for diodes; the presence of a diode at an intersection point enables the signals on the diode inputs—the signals on the corre-

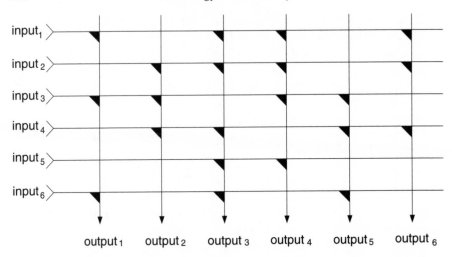

input₁
input₂
input₃
input₄
input₅
input₆

output₁   output₂   output₃   output₄   output₅   output₆

Fig. 7-1. *A diode matrix circuit.* It consists of a set of horizontal wires which receive input signals intersecting with a set of vertical wires which transmit output signals. A diode, when present at a point of intersection, enables a signal to the diode input to be transmitted to the diode output. Diodes are represented here by black triangles.

sponding horizontal lines—to be "gated" or transmitted to the diode outputs— the corresponding vertical lines. Each vertical line will transmit a signal whenever there is an input on any one of the horizontal lines to which it is coupled by means of a diode.

Such a circuit is a far cry from the "random" or "ad hoc" circuits then in use, the latter being not only irregular in structure (and hence complex) but also scaling up poorly. And yet, the idea of the diode matrix forms one of the two central elements in Wilkes's eventual solution. How can this crucial act of ideation be explained in knowledge-level terms?

The records indicate that Wilkes already had prior experience with a diode-matrix-like structure in the case of a component of the EDSAC computer he had just completed—though this component was not part of the computer's control unit.[67] His writings suggest, however, that he did not quite know *how* the diode matrix would serve this purpose. In the summer of 1950, on a visit to the United States, he saw the Whirlwind computer then under construction at the Massachusetts Institute of Technology.[68] In this machine, the execution of each arithmetic operation (except for multiplication, for which a separate scheme was designed) was effected by a sequence of exactly eight pulses issued from a diode "control" matrix. Wilkes's observation of this control matrix, thus, both confirmed his hunch and provided further clues as to how a diode matrix could be employed.

Thus, it seems reasonable to suggest, as was done in *Creativity in Invention and Design*, that at some point Wilkes's knowledge body contained the following "facts," directly gleaned from his EDSAC experiences and his observation of the Whirlwind.[69]

F_4 The EDSAC order interpreter (which issues control signals to the arithmetic, I/O, and memory units) is partially implemented in the form of a diode matrix.

F_5 In the MIT Whirlwind, each arithmetic operation (barring multiplication) involves exactly eight pulses issued from a control unit implemented in the form of a diode matrix.

F_6 EDSAC's order interpreter is simple, repetitive, and regular.

F_7 Whirlwind's control unit is simple, repetitive, and regular.

F_8 The EDSAC order interpreter is a-kind-of order interpreter.

F_9 An order interpreter is a-kind-of computer component.

A series of *inferences* can then be postulated as follows: First, drawing upon F_4, F_5, and F_6 as premises, a *generalization* yields, as a conclusion, a rule

R_2 **IF**   an order interpreter is (partially) implemented using a diode matrix
        THEN   (that part of) the order interpreter is simple, repetitive, and regular.

Next, using rule R_2 and fact F_9 as premises, a further *generalization* of R_2 produces, as a conclusion, the rule

$R_{2.1}$ **IF**   a computer component is implemented using a diode matrix
          THEN   the component is simple, repetitive and regular.

Next, given the empirical fact

F_{11} A control unit is a-kind-of-computer component.

an *instantiation* of $R_{2.1}$ using this rule and F_{11} as premises yields the specialized rule

$R_{2.1.1}$ **IF**   a control unit is implemented in a diode matrix
            THEN   the control unit is simple, repetitive, and regular.

In fact, this rule, which suggests a connection between the properties simplicity, repetitiveness, and regularity of the control unit on the one hand, and the use of a diode matrix on the other, may well have been based on Wilkes's experience with and knowledge of the EDSAC order interpreter alone—as his retrospective remarks about the influence of the EDSAC diode matrix on his thinking support.[70] Rule $R_{2.1.1}$ may then have been further *corroborated* by his observation of the Whirlwind, that is, by facts F_5 and F_7 and the additional observation

F_{10} The Whirlwind control unit is a-kind-of a control unit.

Finally goal G_1 and rule $R_{2.1.1}$ serving as premises can produce, by a kind of inference called *abduction*,[71] the new goal

G_2 Design a control unit to be implemented in terms of a diode matrix.

The cognitive episode just postulated exhibits many of the features I described, in earlier chapters, of the technologist as a cognitive agent and of knowledge-level processes in general. First, it involves a variety of knowledge tokens—facts—specific to the domain of computer design. Second, it entails a structured set (in, this case, a sequence) of actions, each of which involves retrieval of facts and rules (presumably by association and spreading activation) and the execution of an inference resulting in an output—a rule or a goal. Third, this particular (sub) process begins with a goal, and, in this specific sense, is a *rational* process. However, the process itself does not distinguish between conscious and unconscious actions. Whether the inferences by generalization and abduction were done consciously with the agent's full awareness, or were performed in the unconscious state during some incubation state à la Wallas, is irrelevant. *What matters is that the process is explicable as a knowledge-level phenomenon.*

Critical to the plausibility of this explanation is the use of two generalizations, an instantiation and an abduction as means of drawing inferences. This implies that in addition to domain-specific knowledge tokens, such as facts F_5, F_6, and so on, Wilkes's knowledge body is postulated to contain more-abstract rules allowing inferences to be drawn as required. For instance, one of the rules of generalization is

R.*Gen*1   **IF**   an entity e is a-kind-of entity E
              and
              there exists properties P_1 and P_2 such that e satisfies both P_1
              and P_2
              and
              there is no known entity e' that is also a kind of E such that e'
              satisfies P_1 but not P_2
              and
              the variability of P_2 with respect to P_1 is known to be low
         THEN   assume the rule "**IF** E satisfies P_1 **THEN** E satisfies P_2."

This rule specifies conditions under which one may generalize based on the observation of only a *single* phenomenon. The issue of what grounds there may be for a single instance to suffice for the purpose of inductive generalization was raised by the Victorian logician and philosopher John Stuart Mill.[72] In rather more recent times, it was addressed by John Holland and his coauthors[73] and by Paul Thagard, who noted that the validity of such a generalization as "All A are B" as a cognitive act depends on a number of conditions:[74] First, the number of instances of the *co-occurrence* of A and B; second, the *lack of counter evidence*—that is, whether or not there are no observed instances of A being the case but not *B*; and third, the characteristic of *invariability*, in the sense that for the types of events or entities being considered, the extent to which the variability of *B* with respect to *A* is known to be high or low. The "IF" part of the rule *R.Gen*1 embodies these conditions and, if met, allows one to generalize. This rule was applied to produce the rule R_2. Here, the particular, singular entity *e* is the EDSAC order interpreter, the general entity *E* is the class of devices, "order interpreter," the properties P_1 and P_2 are, respectively, "is a diode matrix" and "is simple, repetitive, and regular." The generalization is made possible, furthermore, by the invariability condition being satisfied since the variability of the property "simple, repetitive, regular" with respect to the property "is a diode matrix" was, and is, practically nil. The very nature of diode matrices is such that they would be simple, repetitive, and regular in structure.

The last inference in this episode is that of abduction, and the corresponding rule here can be stated as

R.Abd **IF**   *P* is a problem to be solved (or a goal to be achieved)
                and
                there exists a rule "**IF** the proposition *H* is the case (or the subgoal *H* is achieved) **THEN** the problem (goal) *P* is solved (achieved)"
    THEN   assume that the proposition *H* is the case (or the goal *H* is to be achieved).

Identification of a Stored-Program-Computer-like Form

If the diode matrix constitutes one of the central elements of this ideation, then the concept of the control unit as being a kind of programmed computer in its own right—a computer within a computer, as it were—is the other significant idea central to the overall invention. Indeed, this was the reason for Wilkes's coinage of the word *microprogramming*; it emphasizes the programlike nature of control logic.

This subprocess began with an additional initial goal that stemmed from the fact that a computer's control unit must be *flexible*, in the sense that the sequence of control signals issued by the unit in order to perform some task or operation will *vary* because different operations or tasks demand very different and possibly alterable sequences of control signals. Indeed, it was Wilkes's awareness of the need for this flexibility that prevented him from using the Whirlwind control matrix directly, since it issues a *fixed* sequence of control signals for each arithmetic operation. The following goal, then, can be postulated as an input to this subprocess:

G_3 The control unit must be functionally flexible.

and the output was the (sub) goal

G_4 Specifying the control unit in a stored-program-computer-like form.

As in the previous phase, the inferences used included generalization, instantiation, and abduction.

At this stage, the two key elements underlying the invention of microprogramming have been generated: the idea of a diode matrix in order to achieve a regular structure; and the idea of a stored-program-like form for the sake of functional flexibility. Both these ideations are explicable as knowledge-level processes, as illustrated by stage 3 above. The remaining stages of the entire process describe how the combining or bisociation of these two unrelated ideas can also be explained as knowledge-level processes, entailing exactly the same kinds of goals, knowledge, and inferences.

It is important to reiterate a point emphasized already, that the plausibility of such an account hinges critically on the validity of the knowledge tokens ascribed to the agent. This is why in *Creativity in Invention and Design*, where the knowledge-level process for Wilkes's invention was presented in great detail, much effort was expended in justifying the rules and facts that were drawn upon. Such justification demands intimate knowledge, on the part of the explainer, not only of the technological background of and the public knowledge surrounding the invention, but also of the creative agent's own background and experience. Such knowledge about the agent's knowledge should furthermore be grounded in publicly accessible sources, so that others can assess the validity of the justification advanced. All this makes the construction of plausible knowledge-level processes a formidable enterprise.

Keeping this caveat squarely in one's mind, it is yet tempting to inquire whether any of the other case studies that are woven into this narrative can be explored in the light of KLHI. Can we, for example, shed further light on the thinking that went into the more distinctive aspects of the Britannia Bridge? Do the public records on this project allow us to probe, for example, into how Stephenson may have conceived the critical original idea of the tubular bridge? The section below deals with these questions.

Robert Stephenson's Conception of the Tubular Bridge

We have already seen (in Chapter 6) how the overall design process leading to the mature form of the Britannia Bridge can be explained in terms of the hypothesis law. The very first of these hypotheses was identified as

H_1 A wrought-iron tubular bridge with suspension chains will satisfy the strength, stability, and navigational requirements demanded of the bridge.

How did Stephenson conceive this initial embryonic idea of the tubular bridge? While discussing the variation-selection model earlier in this chapter, I pointed out that this hypothesis was the outcome of the bisociation of three relatively unrelated ideas: the use of riveted wrought-iron plates, the wooden truss, and the suspension chain. Clearly, the formation of this hypothesis was a crucial—perhaps *the* crucial—act of ideation in the entire enterprise. Let us see whether we can sketch the outline of a knowledge-level process describing how Stephenson may have conceived this idea.

Stephenson's personal account records that, initially, he had thought of a cast-iron-arch bridge, this being one of the two main kinds of long-span bridges then in use—the other, of course, being the suspension bridge. However, the arch bridge form had been firmly rejected by the Admiralty because of navigational considerations.[75]

Stephenson's knowledge body would clearly have contained various facts pertaining to suspension bridges, including the recognition that they constituted an important established bridge form and that such bridges were not sufficiently rigid for supporting the movement of trains. Let us codify these as facts.

F_0 The suspension bridge form is a significant long-span bridge form.

F_1 Suspension bridges are not sufficiently rigid for the support of rapidly moving railway trains.

Thus, given a goal, say G_0, to design a long-span railway bridge over the Menai Strait that would satisfy navigational and strength requirements, spreading activation through Stephenson's knowledge body could, we surmise, retrieve fact F_0 and then F_1.

When the arch bridge was rejected, Stephenson felt it necessary to reconsider the viability of the suspension bridge form with additional stiffening of the deck.[76]

We have here an expression and application of what must be regarded as a fundamental engineering creed: dissatisfaction with the way things are and the consequent desire to rectify or improve the state of affairs (see Chapter 3).

In fact, this can be stated as one of the very general rules that would secure a place in every engineer's knowledge body:

R_0 **IF**   an artifact demonstrates some negative or disadvantageous character
       THEN   establish as a goal the elimination of that character.[77]

Stephenson's original goal, once the cast-iron-arch form was rejected, may have suggested the suspension bridge form, which in turn activated fact F_1; fact F_1, expressing a negative character of an artifact could, by spreading activation, retrieve R_0. We can now see that an inference, by the rule of inference called *modus ponens*, conducted consciously or unconsciously by Stephenson with F_1 and R_0 as premises, may have led to the conclusion embodied in the goal[78]

G_1 Eliminate the lack of rigidity of the suspension bridge form so that it can support rapidly moving trains.

We have no indication from Stephenson's personal account, or from those by Edwin Clark and William Fairbairn, whether he considered alternative approaches for stiffening the bridge form—whether, in fact, variations were generated. What we do know is that, at about this time, his contemporary J. M.

Rendel called his attention to a method of trussing perfected by Rendel for preventing oscillation in the platform of suspension bridges.[79] Rendel's system had stood the test of time for several years, and had, apparently, been reported in the *Transactions of the Institution of Civil Engineers* in 1841. Thus, a new set of tokens can be said to have been introduced into Stephenson's knowledge body, these pertaining to Rendel's trussing system. It is reasonable to assert that among the relevant tokens is the simple fact

F_2 Rendel's trussing system prevents oscillation in suspension bridges.

However, Stephenson felt that while Rendel's trussing system was wholly satisfactory in the case of road bridges—it had been employed by Rendel to build one called the Montrose Bridge—it would not be adequate to support the weight of passing trains. The latter demanded "a much stronger and more ponderous system" than was used for the Montrose Bridge.[80]

It would appear, then, that Stephenson either knew that Rendel's system, as it had been used by the latter on the Montrose Bridge, was insufficiently strong for railway purposes or (more likely) that he reasoned out this conclusion. Let us see how such a reasoning could be described in knowledge-level terms.

Rendel's trussing system was, apparently, employed for the Montrose Bridge; the latter was an instance of, or a kind of, suspension bridge; the trussing established an adequate level of rigidity for the purpose at hand. These were all empirical facts. Suppose we use the following notation to represent the above:

   e: the entity Montrose Bridge
   E: the entity "suspension bridge form"
   P_1: the use of Rendel's trussing system
   P_2: the establishment of rigidity or lack of oscillation

In that case, we can also assert the facts that

e is-a-kind of E
e satisfies P_1
e satisfies P_2

In that case, assuming that there were no other instances of bridges built according to Rendel's trussing system that were not rigid or that exhibited oscillation, a *generalization* may be postulated producing the rule

R_1 **IF**   Rendel's trussing system is used for suspension bridges
        THEN   rigidity of the suspension bridge will obtain.

Using goal G_1 and rule R_1 as premises, an inference by *abduction* will result in a new (sub) goal as conclusion:

G_2 Apply Rendel's trussing system to the Britannia Bridge problem.

And yet, as the quote above from Stephenson's account indicates, he was not convinced that the goal G_2, if achieved, would in fact result in the attainment of its parent goal G_1. The argument toward this conclusion can be explained as follows:

Let us suppose the existence, in Stephenson's knowledge body, of the following uncontroversial facts:

F_3 The Montrose Bridge is a-kind-of road bridge.

F_4 The Britannia Bridge is (to be) a-kind-of railway bridge.

F_5 Rendel's trussing system was the structural style for the Montrose Bridge.

F_6 Road bridges are not generally as strong as railway bridges.

That such facts may be present in Stephenson's knowledge body need very little justification. Furthermore, by spread of activation beginning with goal G_2—which is a goal about Rendel's trussing system—these facts can be associatively retrieved. Let us further suppose the presence, in Stephenson's knowledge body, of the following token, which is specific to the domain of structural and mechanical engineering:

R_2 **IF** e_1 is a-kind-of structural entity E_1
            and
            e_2 is a-kind-of structural entity E_2
            and
            E_1 is generally less strong than E_2
            and
            e_1 is designed according to structural style S_1
       THEN   there is no guarantee that e_2 can be designed according to S_1.

I have no evidence at hand that Stephenson actually ascribed to this belief. *If* he did, however, then facts F_3, F_4, F_5, F_6 would match the **IF** part of R_2 and *modus ponens* would allow us to deduce

F_7 There is no guarantee that Rendel's truss is sufficiently strong for the Britannia Bridge (suspension bridge) form.

This gives us an idea how Stephenson's conclusion that "a much stronger and more ponderous system" than was used for the Montrose Bridge can be depicted as a sequence of knowledge-level actions. The goal G_2 was, according to this account, abandoned but not the idea of using trussing. For, after considering "a variety of devices" for "a stronger and more ponderous system"—regrettably, we are not told what these variations were—Stephenson hit upon the idea of combining the suspension chain with trellis trussing, the latter forming the vertical sides and attached by means of suspension rods to the chain. The sides would, furthermore, be held in position by cross-braced frames at the top and bottom. In effect, what this resulted in was "a roadway surrounded on all sides by strongly trussed framework."[81]

Here is the first crucial step of bisociation! Now, it is certainly possible that the structure described above was reasoned out by a generalization of the Rendel structure—a generalization in which the trussing of the vertical sides of the bridge were heightened and then retained with the use of cross bracing at the top and bottom. But Stephenson also alludes to the fact that a structure of this sort had been used in the United States to build a canal aqueduct and had been described in the forty-fourth volume of the *Mechanics' Magazine* for 1846.[82]

His account does not make it at all explicit as to whether he was actually aware of the American aqueduct in 1845 at the time he began work on the project. But he does go on to say: "The application, however, of this system to an aqueduct is perhaps one of the most favorable possible; for there the weight is constant and uniformly distributed, and all the strains consequently fixed both in amount and direction, two important conditions in wooden trussing constructed of numerous points. In a large railway bridge . . . these conditions . . . are ever varying to a very large extent but when connected with a chain . . . the direction and amount of the complicated strains throughout the trussing become incalculable."[83]

The tone of this passage seems to suggest that he may have been aware of the American aqueduct—in which case the *idea* of the tubelike truss was obtained directly from this knowledge.

This *borrowed* idea can be represented by the facts

F_8 There is a canal aqueduct designed and built in the U.S. that consists of wooden trussed sides and cross-braced top and bottom.

F_0 The American canal aqueduct is strong.

From this a *generalization* can lead to the rule

R_3 **IF**   the goal is to make a long, strong bridgelike structure
      THEN   design the structure with deep trussed sides and cross-braced top and bottom

A further inference by the rule of *modus ponens*, using goal G_1 and rule R_3 as premises, yields as conclusion, the following goal:

G_3 Design the suspension bridge with deep wooden trussed sides and cross-braced top and bottom.

This, incidentally, can be stated as a *hypothesis* about the bridge design that precedes the hypothesis H_1:

H_0 A suspension bridge with deep wooden trussed sides and cross-braced top and bottom will satisfy the strength, stability, and navigational requirements demanded of the bridge.

But if Stephenson did indeed borrow the idea from the American example, he then went on to criticize it, as the last quotation above suggests.

Such criticism can be effected by appealing to a very general problem-solving strategy, first identified in 1972 by Allen Newell and Herbert Simon, called the *means-end heuristic*.[84] This strategy, suitably tailored for the specific realm of design, can be stated in rule form as follows:

IF there exists a goal G specifying a requirement for an artifact

   and

   there exists a design D intended to satisfy G

THEN determine the misfit between D and G

   and

   then apply an action to eliminate or reduce the misfit.[85]

I shall postulate that Stephenson's criticism of the scheme described in hypothesis H_0 (or, equivalently, goal G_3) entailed the application of this rule. Its application to H_0 with respect to the original goal G_1 may yield several misfits including two that Stephenson mentions: (a) "the direction and amount of the complicated strains throughout the trussing becomes incalculable" and, hence, whether the requisite strength is obtained is, presumably, indeterminable; and (b) the "perishable nature" of the wooden truss and its vulnerability to fire.

It was then that, in order to eliminate these misfits, Stephenson decided to use wrought-iron riveted plates rather than wooden trussing.

Here again, he drew upon a previous design of his own for a small bridge made in 1841. This bridge was constructed of wrought-iron plates riveted together but otherwise "arranged as an ordinary cast-iron girder bridge." Recalling this structure, Stephenson was led to the idea of applying wrought iron as a means of stiffening the platform of a suspension bridge. His initial solution was, then, to substitute the vertical wooden trellis trussing and the top and bottom cross braces with wrought-iron plates riveted together. The resulting form was "a huge wrought-iron rectangular tube, so large that railway trains might pass through it with suspension chains on each side.[86]

The reasoning is virtually laid bare here. To eliminate the misfits identified, Stephenson retrieved, from his knowledge body, an earlier design that had a stronger structural basis, namely, wrought-iron riveted plates; he then substituted the wooden form with the equivalent wrought-iron form. We can postulate that such substitution may have come about as a result of using an analogy. The outcome was the initial version of the tubular bridge as described in hypothesis H_1.

If we consider the act of design or invention in the large, and we accept that the knowledge-level model is a very useful cognitive theory of the design process, we see that both the hypothesis law and KLHI, the knowledge-level hypothesis on ideation, play roles in the overall scheme of things. The development of an artifactual form from an original goal to a validated mature state unfolds according to the hypothesis law. Each hypothesis in itself may be regarded as the outcome of an act of ideation that can be described as a knowledge-level process. In fact such ideation when examined more minutely may itself entail the construction of transient hypotheses, as we saw in Stephenson's arrival at the initial form of the Britannia Bridge.

We have also seen that the variation-selection mechanism has a role in this picture: At any given stage of the knowledge-level process, alternative hypotheses may be generated from which a selection is made. Variation-selection, at least as revealed at the knowledge level, is not, however, an *essential* component of the creative process. Finally, we have seen how the mysterious act of illumination, à la Graham Wallas and his distinguished predecessors and bisociation in Arthur Koestler's famous sense are explicable in terms of the knowledge level.

Suppose now that we were to construct a computer simulation of any of the technological episodes discussed thus far. Many issues that are simply invisible at the knowledge level would have to be dealt with in order to implement the simulation—for instance, how to represent and encode knowledge. But one of the cognitive features that *is* visible, at least implicitly, at the knowledge level and would definitely have to be considered explicitly is that of *search*. Let us, then, consider the phenomenon of search in the specific context of technological ideation.

The Role of Search in Ideation

Search is, of course, inevitable once we accept bounded rationality. If we are unable to take the *exact* action or sequence of actions needed to achieve a goal, all we *can* do is tentatively try out actions—to search for a solution or a way of achieving the goal. The whole idea of the hypothesis law rests on this notion, at least tacitly.

Search is manifested, in the course of ideation, in two different ways. First, given a goal, the agent needs to search for facts, rules, and other tokens in its knowledge body that will allow him or her to take appropriate action toward the attainment of the goal. Thus, one form of search is that of the knowledge body itself. Following Allen Newell, I shall call this *knowledge search*.[87]

Second, given a goal, the agent will conduct a search for a solution in a space which Newell and Simon in their book *Human Problem Solving*, called the *problem space*.[88] Newell also refers to this form of search, because it involves the exploration of a problem space, as *problem search*.[89] He has furthermore pointed out that these two kinds of search differ in a rather fundamental way: Knowledge search occurs primarily in an existing structure (of knowledge tokens) whereas problem search *constructs the very space that is to be searched*.

Now, the latter sounds very mysterious. We all intuitively understand the general nature of knowledge search—because in many circumstances we do it quite consciously—as, for example, in searching for a word in a dictionary, or for a book in a library. But problem search seems more problematic. What does it mean for a "space" to be "constructed"?

In a certain sense, the word *space* is used as a *metaphor*: It allows us to relate a poorly understood phenomenon to something we do understand—or, at any rate, we are familiar with.[90] Thus, to say that in the course of deriving a solution to a problem, say a design problem, we conduct a search through some space, is to invoke, as metaphor, the idea of search through physical space, as in negotiating a maze or seeking a tractable path up a hilly terrain. The metaphor of space serves as a visual aid for our mind's eye.

And yet, the metaphor of physical space falls short in that physical space is already there before a physical search begins. The maze is there to be explored, the hill to be climbed. Physical search *discovers* the space, as it were, whereas in the case of problem search, one speaks of *constructing* the space in which to search.

To understand more clearly the nature of problem search, let us then explore further this notion in the specific context of technological ideation. For this purpose, I will sketch out a small part of a design process that is far less creative that those described earlier in this chapter. The process as described may be classified, at best, as being PN-creative. However, it will admirably suit our purpose.

The Nature of Problem Search in the Design Process

Let us consider a design problem from the realm of high-performance computer systems. In particular, the problem at hand is the design of a *parallel file system*. This is the target artifact. It will consist of both a hardware subsystem to perform the physical input and output (I/O) functions of the computer, and a file-processing software subsystem. The two, in combination, create what computer systems engineers would call a "virtual" file-processing machine. The designer's task is to describe the detailed architecture of this parallel file-processing machine.

Let us suppose that we are somehow in a position to observe the systems engineer as she performs this task, and are privy to her thoughts in the course of the design. This could be effected by having her verbalize her thoughts and recording the resulting protocol, as Elaine Kant and Allen Newell did in the case of the convex-hull algorithm design experiment described in Chapter 6. Or it could be achieved if the agent were to follow an explicit design method which forced her to record her thoughts and design decisions.[91] We shall, accordingly, follow the systems engineer, at least in the very first stages of the process, to see what it means to conduct search in a problem space—which in the current, specialized context we may refer to as a *design space*.

The target artifact is, to repeat, a parallel file system. Let us name this artifact simply PFS. The requirements for the system have been identified. They are many in number, and unfortunately they are not all mutually compatible or consistent. Requirements, in general, may interfere with one another. Inevitably, as the design proceeds, some of the requirements may have to be sacrificed or, at the very least, relaxed to some extent. However, at the very early stage of the design process, the designer, because of bounded rationality, may not even comprehend the manner in which a given set of requirements may interact. In the initial stage of design, she may quite freely pursue the problem as if the requirements are independent—until and unless she has explicit knowledge to the contrary. Eventually, of course, a stage will be reached when interferences and incompatibilities will have to be resolved.

Accordingly, we shall follow the designer as she pursues just one of the initial requirements identified for PFS. This requirement, stated in the form of a goal, is

G_0 PFS will be a high-performance system.

This goal, in effect, establishes, the initial problem space in which the designer plans to work. It consists of the set of all conceivable parallel file systems that satisfy the property of yielding a high performance.

In considering how to achieve G_0, the agent draws upon her knowledge body and retrieves several tokens that pertain to high-performance systems. Details of these tokens are unimportant in the present context, as our concern is with the nature of the problem search. But, as in the case of the knowledge-level processes described earlier in this chapter, the tokens are applied by the agent to infer the following conjunctive subgoals for G_0:

G_1 PFS will provide high system bandwidth.

G_2 PFS will provide low I/O latency.

G_3 PFS will support high system throughput.

We have here a first glimpse of problem search in the design space. The original goal G_0 can be met by a vast range of possible designs. This is because the property "high performance" can be met in a variety of ways depending on one's views or preferences. So what the designer does is to *reduce* the space of potential solutions by establishing more precise criteria for high performance— namely, high bandwidth and throughput, and low latency.[92] Each of these requirements or goals defines a subspace of the initial space. And since the goals G_1, G_2, G_3 are conjunctive relative to G_0, all three must be met. The resulting subspace to search has, thus, been reduced to a space that is *common* to the three subspaces.

The designer next decides to determine how goal G_1 can be achieved. Using rules pertaining to bandwidth, she establishes as subgoals for G_1 the following conjunctive goals:

G_4 PFS will have a high maximum bandwidth.

G_5 PFS will have a high sustained bandwidth.

Here "maximum bandwidth" refers to the maximum amount of data transfer per unit time that can be achieved by the target artifact. It is a measure of the peak performance of the system under the most ideal operating conditions. "Sustained bandwidth," in contrast, signifies the average amount of data transfer per unit time for a typical or "characteristic" workload that may be expected of the artifact. It is, thus, a more realistic measure of bandwidth performance.

The reader will realize that G_4 and G_5 individually define subspaces of the space defined by G_1. And since G_4 and G_5 are conjunctive subgoals for G_1, the subspace of interest is, thus, that which is common to G_4 and G_5. Assuming, in general, that some part of this common subspace is also common to goals G_2 and G_3—at this point, there is no evidence to indicate otherwise—the candidate space has been further reduced.

Let us skip the next few steps and review the situation somewhat later in the design process. By this time, suppose that the designer has drawn upon her knowledge body to propose, as an initial version of PFS, a hardware-software *model* system that had been previously published in the technical literature (Figure 7-2). This model may be referred to as MODEL-PFS. It specifies the overall structural form of the file system in terms of its major functional components and their relationship to one another, but nothing is stated regarding specific details as to the number of components, their performance characteristics, and so on. These latter elements constitute the target artifact's parameters.

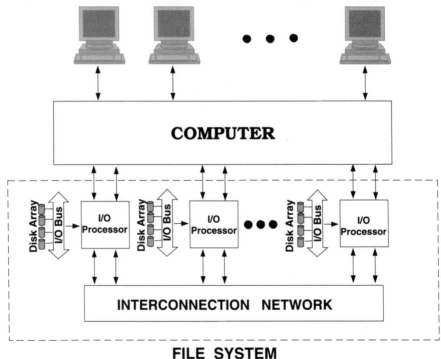

Fig. 7-2. *A model parallel file system.* A set of input-output (I/O) processing devices are connected to a computer and to one another by means of an interconnection network. Each I/O processor controls an array of disks. The software controlling the hardware enables the system as a whole to operate and process data files in a parallel fashion.

What is significant about this state of affairs, as far as problem search is concerned, is that the designer has substituted what Newell called "immediate or prepared" knowledge for deliberation,[93] and thereby both added a new subspace and *reduced further the size of the effective candidate space*. For, just as the requirements stipulated in the form of a goal determine the space of all designs that satisfy the requirements, so also a conceptual, structural form of the artifact with the parameter values missing determines a space of all designs that have this same structure but with different parameter values. The conceptual design MODEL-PFS thus serves the same function as the goals in identifying a candidate space for the target artifact.

The design process just sketched, though brief, gives a fairly explicit idea about the nature of problem search. A few points are worth commenting on in this regard.

First, I think it is misleading to assume or claim that the designer, in attempting to derive a design for the parallel file system, was *consciously* searching some problem space. In the scenario I have just painted—and I think this is also true for the more creative sorts of ideation discussed earlier in the chapter—the agent does not say, "Let me construct an appropriate problem space and search for a solution in that space." For the designer, the problem is defined by the goals at hand, and the knowledge available to her. In attempting to achieve her goals, our engineer invoked certain knowledge tokens and performed certain actions that led to the generation of subgoals or, eventually, to the postulation of a structural form for the target artifact. The *effect* of such a knowledge-level process was the creation of problem (or, in this context, design) spaces, and the narrowing down of the space in which the solution was thought to reside. The designer did not consciously search for a solution, but she acted *as if* she was conducting a problem search.

Second, the example helps to shed some light on Allen Newell's somewhat cryptic comment that "problem search constructs its own space."[94] In the case of technological ideation—which is what we are interested in here—a goal is a way of specifying one or more requirements that the intended artifact is to satisfy. To achieve such a goal is to specify the form or structure of an artifact that meets the corresponding requirements. Thus, a goal is a succinct way of specifying a design or a (possibly unbounded) family of designs that meets the requirements mentioned in the goal. It is this set of designs corresponding to the statement of a goal that constitutes a design space. And since a design process, being a knowledge-level process, entails the progressive generation of goals, each such goal generated also generates, in effect, a space of artifactual forms or designs, much as an algebraic equation such as $y = 3x + 4$ is a succinct definition of and can generate an infinite number of points on a straight line. As we have seen from the brief example of the file system, the dynamics of the design process as represented by the changing nature of the goal set is mirrored by the construction and interaction of design spaces.

Finally, it will be noticed that the shape of the problem search process is dictated by the knowledge the agent uses in the course of ideation. It is the knowledge that the agent brings to bear on the problem at hand that determines the extent of the search. We have encountered numerous instances throughout this book of how an agent draws upon certain key aspects of his knowledge body to enhance the process of design or invention; to hasten the search, in effect. The file system designer introduced a previously proposed design, thereby substituting "immediate" or, as it is also called, "compiled" knowledge for the effort of deliberation or more ponderous search. Robert Stephenson invoked his experiential knowledge of suspension bridges and application of riveted wrought-iron plates, and combined these with acquired knowledge of Rendel's trussing system, in the course of his quest for a form for the Britannia Bridge. Maurice Wilkes drew upon his knowledge of the diode matrix and its application in both the EDSAC and the Whirlwind to achieve the regularity he desired in the control unit. The Mond metallurgists, seeking their high-strength creep-resistant alloy, vastly reduced their search by utilizing their shared knowledge of the properties of certain nickel and chromium-based alloys. Alexander Graham Bell's knowledge of the structure and function of the human ear, and of Helmholtz's apparatus for producing vowel sounds was, literally, instrumental in shaping and delineating the process that led eventually to the telephone. Indeed, insofar as the examples illustrate the notion of bisociation, it is obvious that for bisociation to happen demands knowledge in the first place. As a final example, Benjamin Huntsman's invention of "cast steel" by what came to be known as the crucible process demanded that raw steel be first melted down. This idea was apparently derived from Huntsman's knowledge of *brass founding*, which entailed precisely such an activity.[95] Ideation, even of the most creative kind, is knowledge intensive.[96]

8

Creativity and the Evolution
of Artifactual Forms

Why, I asked at the beginning of this book, had the blast furnace taken the form it had? What "stroke of genius" led the inventor(s) of this gigantic artifact to conceive its particular form and not some other? I could have, of course, asked this same question with respect to any other artifact; about Maurice Wilkes's invention of the microprogrammed control unit, or of Stephenson's tubular bridge, or of the Mond metallurgists' high-strength creep-resistant alloys. In these specific cases, I attempted to provide some answers. And in so doing, it became clear that the act of creation is a knowledge-intensive enterprise. The creative technologist draws upon a diversity of knowledge tokens during the process of ideation.

In the next chapter, I will dwell on another of the issues raised in the Prolegomenon—namely, the nature of technological knowledge and its relationship to what we call science. For the present, however, let us consider one specific kind of knowledge that pervades technology: the knowledge of past, possibly even vanished artifacts that the inventor or the designer brings to bear in conceiving and creating a new artifact. Wilkes drew upon his knowledge of the MIT Whirlwind's control unit. The Mond metallurgists began their search for creep-resistant alloys for gas-turbine blades with a family of nickel-chromium alloys. Robert Stephenson's tubular bridge in its early form had the vestiges of both the suspension and the truss bridges.

These examples seem to tell us that the past is, in one way or another, present in acts of technological creation. That is, in order to explain how some particular artifactual idea or form is born in an agent's mind, it may be necessary to invoke the past. History, contrary to what Henry Ford is alleged to have said, is far from bunk, when it comes to explaining creativity. Even the most original product of the technologists's mind—what I have referred to as historically original—is, so to speak, a product of history itself. If we examine the history of an individual artifact we may discover that it exhibits a phenomenon that we can only describe as *evolutionary*.

Since the word *evolution* carries with it a lot of baggage, it is necessary to explain precisely what I mean when I say that the history of an artifact is evolutionary.

122

We have already seen in Chapter 7 how certain Darwinian notions, notably, variation generation and selection therefrom, have been applied to explain the process of ideation. Variation-selection is, of course, one facet—and a strictly Darwinian facet—of evolution. But we have also seen, in Chapter 6, that on a longer timescale, during which a single mature artifact comes into being, the process entails the formation of a succession of hypotheses, each resulting from a modification of an earlier one. This too is evolution, though not of the Darwinian variety. Rather, one might liken the evolution of an artifactual design from its original set of requirements through various stages of intermediate and incomplete forms to a mature form—as illustrated, for example, by the development of the Britannia Bridge design—to the biological concept of *ontogeny*, a word that refers to the "life history of an individual both embryonic and postnatal."[1] Thus, the process whereby an individual artifactual design comes into being we can call *ontogenic evolution*.[2] The hypothesis law of maturation is an expression of ontogenic evolution.

The concern of this chapter, however, is with a still longer term phenomenon, involving years, decades, and in some cases even centuries; not the history of an individual artifact but the *evolutionary history of a linked network of mature artifacts or forms leading up to the invention of a given artifact of interest*. Taking another leaf from the biologists's book, we can name this *phylogenic evolution*[3] since *phylogeny* in the organic world refers to the "evolutionary history of a lineage conventionally . . . depicted as a sequence of successive adult stages."[4]

That artifacts have evolutionary histories (in the phylogenic sense) is by no means a new idea. In his fascinating book *The Evolution of Designs*, the design theorist and architect Philip Steadman has described, albeit critically, various types of biological analogies that have been applied to architecture and the decorative arts.[5]

Perhaps the best-known instance of the application of the evolutionary idea is the work of the nineteenth-century British soldier and ethnologist Lieutenant General Lane-Fox Pitt-Rivers, who organized, arranged, and displayed his collections of preindustrial artifacts as sequences of closely related forms.[6]

The most systematic study known to this writer of the evolutionary nature of technology as a whole is that of the historian George Basalla, who in *The Evolution of Technology* sought to explain technological change within an evolutionary framework.[7] More recently, Henry Petroski in *The Evolution of Useful Things* explains how certain artifacts have come to be the way they are by tracing their evolutionary antecedents.[8]

It is precisely this presence of the past, this assimilation of evolutionary history into an individual agent's knowledge body and its influence on that body, that I wish to examine in this chapter. Very specifically, the questions to be explored here are the following: To what extent are significant acts of technological creativity dependent on what has already happened in the past? To what extent is invention determined by evolutionary history?

As in the case of most of the issues raised in this book, these are empirical questions that demand scrutiny of historical cases. I shall, accordingly, consider two rather significant examples of historically original inventions to see whether in fact we can discern a clear presence of the past in the creation of original artifacts.

The Newcomen Engine

One of the prime movers of the Industrial Revolution—which had its beginnings sometime between the 1760s and 1780s[9]—was the harnessing of steam power. Thus, the development of the steam engine and, in particular, the patenting in 1769 of the separate condenser engine by James Watt (1738–1819) is considered to be what Donald Cardwell would describe as one of the major turning points in Western technology.[10]

But while Watt was and remains the most significant figure in the history of the steam engine, he was by no means its pioneer. Rather, Watt's invention was the outcome of his dissatisfaction with the efficiency of a model version of the Newcomen engine that he had been asked to repair.[11] In the history of technology, it is Thomas Newcomen (1663–1727), an English ironmonger,[12] who is usually credited with the invention of the first continuously operating self-acting engine that utilized steam.[13] That the Newcomen engine was a historically original invention is firmly established. The question of interest is: To what extent did Newcomen owe a debt to what had gone on before?

The records definitely indicate that a Newcomen engine was installed and made fully operational in 1712, although an earlier version may in fact have been built in 1705.[14] In any case, there seems to be some evidence that Newcomen, with the aid of his assistant John Calley, had worked on the development of his engine for a period of ten years or more, beginning around 1698.[15] This date is of some importance to our inquiry, as we shall see.

Figure 8-1 shows the essence of the Newcomen engine as it was installed in 1712. It consists of a large metal cylinder fitted with a piston. The piston is connected by means of a chain to one end of a pivoted beam, which is connected at the other end to a mine pump. The cylinder is filled with steam from the boiler below it, thereby equalizing the pressure of the atmosphere on the cylinder. As a result, the weight of a pump gear at the other end of the beam causes the piston to ascend to the top of the cylinder. When this happens, the supply of steam is cut off, a jet of cold water is sprayed into the cylinder, and the steam condenses leaving a vacuum under the piston. Atmospheric pressure then drives the piston down. When it reaches the bottom of the cylinder, the spray is turned off, the steam supply resumed and the cycle of operation begins again.

If we now wish to claim that Newcomen's invention evolved from earlier inventions that had been developed over a possibly significant period of time— that is, if we wish to claim that the Newcomen engine has an evolutionary history of a phylogenic nature—we must demonstrate that (a) certain key features, components, or aspects of Newcomen's engine were already known or had been previously invented; and (b) Newcomen himself was aware of these previously established components or inventions.

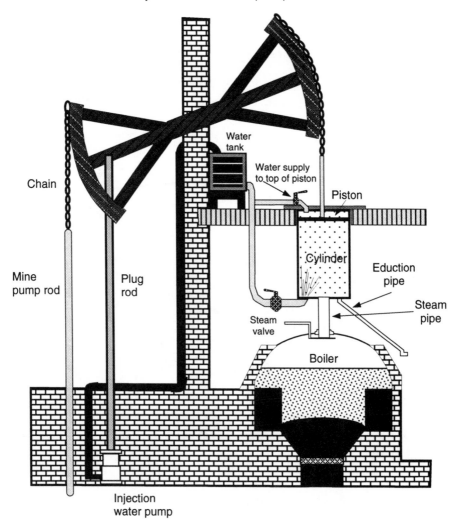

Fig. 8-1. *The Newcomen engine, circa 1712* (Adapted by permission of the publisher from H. W. Dickinson, *A Short History of the Steam Engine,* © 1939, Cambridge University Press). Steam enters the cylinder from the boiler and equalizes the atmospheric pressure on the piston. As a result, the piston rises because of the weight of the pump gear attached to the other end of the beam. The steam supply is cut off, and cold water is sprayed into the cylinder, causing the steam to condense. A vacuum results and the pressure of the atmosphere forces the piston down.

Notice that the assertion that the Newcomen engine had an evolutionary history does not diminish in any way its inventor's creativity. The Newcomen engine was the *first* fully self-acting engine that utilized steam; it was widely employed in England in the eighteenth century; and it led in turn to James Watt's later and more permanent achievement. But if, in fact, we can demonstrate that this engine did indeed have an evolutionary past, then we will have shed some light on the general question: "Why has a given artifact taken the form it has?" For we can then answer: "This form is based on what has been achieved before. This form is based on an earlier form." *That* is the essence of the evolutionary concept.

Fortunately, the history and the prehistory of the steam engine has been documented quite extensively by scholars, and I shall draw freely and gratefully upon such historical material. The story appears to begin with Evangeliste Torricelli (1608–1647), a sometime disciple of Galileo, whose experiments in 1643 led him to conclude that the atmosphere exerted a pressure. Torricelli can be said to have discovered the phenomenon of *atmospheric pressure*, the most significant *scientific* aspect of the engine Newcomen built some seventy years later.

The second significant scientific (rather than technological) discovery of relevance was the fruit of the experiments described by the Dutchman Otto von Guericke (1602–1686) in his *Experimenta nova Magdeburgica de Vacuo Spatio*, published in 1672. This was the *power* of atmospheric pressure. In the most famous of Guericke's experiments, he put two small hemispheres together to form a sphere, and after evacuating the air by means of an air pump, he showed that sixteen horses could not overcome the atmospheric pressure imposed on the sphere and pull the hemispheres apart. In another experiment, Guericke showed that if a piston (using the modern term) is inserted into a cylinder and the latter evacuated then teams of men were unable to pull the piston out against the pressure of the atmosphere. Perhaps most significantly, in the present context, Guericke actually hinted at harnessing the force of the atmosphere for practical purposes.[16]

In the same year that Guericke's *Experimenta* was published, his compatriot, the astronomer Christian Huygens (1629–1695) proposed an alternative method to using the air pump for producing a vacuum. He suggested exploding a small charge of gunpowder in a cylinder closed by a piston that would expel the air in the cylinder through a valve in the piston. However, it was not easy to produce a good vacuum in this way. There was also the difficulty of igniting the gunpowder.[17]

The next important element in the story is the work of Denys Papin (1647–1712), a French protégé of Huygens who, from 1675 on, worked with Robert Boyle in England on the development of an improved air pump. Two of the new features of this air pump were that the valves opened and closed by the action of the machine itself; and pistons were covered with water to make them airtight.[18] In 1691 Papin conceived a method for producing a vacuum by the condensation of steam in a cylinder under a piston.[19] A pool of water at the bottom of the cylinder is boiled by means of a fire below it; the resulting steam forces the piston up. When the latter reaches the top, the fire is removed. The cylinder cools, the steam condenses, and a vacuum results. The presence of the atmosphere drives the piston down.

The story thus far has been concerned solely with the issues of atmospheric pressure and the production of vacuum. There is another matter of relevance here—the concern in the seventeenth century with the problem of *raising water*.

Both these concerns are important because of their confluence in the Newcomen engine.

In 1615 a French landscape gardener to various courts of Italy, England, and Germany, one Solomon de Caus, described an apparatus for raising water by steam pressure.[20] This apparatus consisted, simply, of a small spherical copper boiler into which a pipe is inserted reaching almost to the bottom and projecting outside. When the sphere is partly filled with water and heated, the resulting steam forces the remaining water out through the pipe.

In the English court, at the time de Caus was employed there, was a prolific inventor called David Ramsay, who, in H. W. Dickinson's words, "could hardly help having known de Caus."[21] One of Ramsay's patents, issued in 1631, covered a number of inventions including one "To Raise Water from Lowe Pitts by Fire."[22] Attending the English court at the time Ramsay was awarded his patent was Edward Somerset (1601–1667), later the marquis of Worcester. As Dickinson comments, "it can hardly be doubted that he [Worcester] would become acquainted with Ramsay and learn what he was trying to do."[23]

In 1663 Worcester published a book entitled *A Century of the Names and Scantlings of Inventions by me already Practised*. One of these inventions was an engine for raising water by steam—or as Worcester apparently termed it, a "water-commanding engine."[24] There were no drawings accompanying the description; however, a schematic outline of the engine was reconstructed by Worcester's biographer, Dircks, and is shown in Figure 8-2. A version of this engine was apparently installed around 1628 and another improved version built in 1647.[25]

As Figure 8-2 indicates, steam is admitted from the boiler C through a four-way cock b and pipes B, B' into the reservoirs A, A' which are connected with the source S of water by means of the pipes G, G'. The reservoirs are also connected with outlet pipes F, F' that, in turn, meet at a four-way cock a. A partial vacuum is produced in the initially empty reservoir, say A, by the condensation of a part of the steam. This forces water up into the reservoir by the pressure of the atmosphere acting on the surface of the water in S. When A is nearly full, steam is admitted and its pressure forces the water out of the reservoir up through F and into the pipe E. When the reservoir is nearly empty, the steam is shut off resulting in the rapid condensation of the steam and formation of a vacuum. The two reservoirs A and A' work in flip-flop mode: While one is filling with water, steam forces the water in the other up into E.

This description of Worcester's engine, based on the accounts given by Robert Thurston and A. P. Usher[26] indicates clearly the deployment of *both atmospheric and steam pressures* as well as the creation of a vacuum by condensation of steam. As we shall see below, the form and operation of this engine is remarkably similar to the one developed later by Thomas Savory—the latter being the immediate precursor to the Newcomen engine.

In 1682 Sir Samuel Moreland (1625–1695), "Master of Mechanicks" to King Charles II, showed the king "a plain proof of two several or distinct trials of a new invention for raising any quantity of water to any height by the help of fire alone."[27] There are doubts, though, that the machine was ever actually

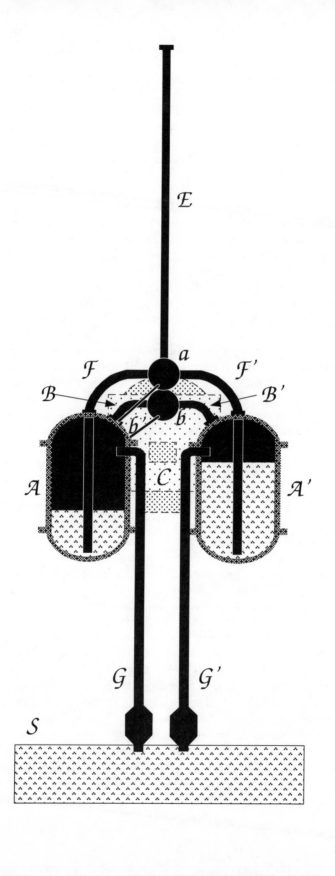

constructed. In 1936, however, the historian H. W. Dickinson identified a sketch of Moreland's machine described in the diary of one Roger North, a lawyer. According to the description accompanying the sketch, it appears that Moreland's machine is the earliest known example of the employment of steam from a separate boiler to exert pressure on a piston in a closed cylinder.[28] It is also, according to R. T. C. Rolt, the oldest known example of an automatic or self-acting valve gear.[29] Robert Thurston has also pointed out, referring to a book written by Moreland called *Elevation des eaux par toute sorte de machines* and published in Paris in 1683, that Moreland showed considerable familiarity with the characteristics of steam power. He even presented a table relating the volume and pressure of saturated steam.[30]

Finally, we come to Thomas Savery (1650–1715), Fellow of the Royal Society, whose "fire engine" was, according to Cardwell, the first to go beyond the paper design or working "toy" stage.[31] Savery's engine was described by its inventor in a pamphlet entitled *The Miner's Friend* published in 1702.[32] However, well before the appearance of this pamphlet, Savory had exhibited a model of his fire engine to King William III in 1698 and had obtained on July 25 of that year a comprehensive patent which granted him the sole exercise of "a new invention for Raiseing of Water, and occasioning Motion to all sorts of Mill-Work, by the Impellant Force of Fire, which will be of great use and Advantage for Drayning Mines, Serveing Towns with Water and for the Working of all Sorts of Mills, when they have not the Benefit of Water nor constant Windes."[33] The patent was to be held for fourteen years.

Savery also submitted a working model to the Royal Society in 1699 and demonstrated its operation at one of its meetings. He supplied the society a drawing and a specification of the engine; the *Philosophical Transactions of the Royal Society* published a copperplate engraving and description of the model.[34]

Basically, Savery's engine (Figure 8-3) consisted of a furnace A, heating a boiler B connected by pipes to two receivers D, D'. Two pipes F, F', connected to the bottom of the receiver, turn upward and join to form a "forcing-pipe" G. A pipe from the top of each receiver turns downwards; the two join to form a suction pipe H leading down to the well or other source from which the water is to be raised.

Steam produced in the boiler passes through a cock C and fills one of the receivers, D. C is closed. The steam condenses in the receiver, a vacuum results,

---

Fig. 8-2. *The Marquis de Worcester's "water-commanding" engine, circa 1628* (Adapted by permission of the publisher from A. P. Usher, *A History of Mechanical Inventing,* © 1982, Dover Publications). Steam enters one of the reservoirs (say A) from the boiler C, condenses, and creates a partial vacuum. The atmospheric pressure on the water source S forces water up into A. When A is nearly full, steam is admitted and forces water out of the reservoir into the pipe E. When A is nearly empty, the steam supply is cut off, the steam in A condenses, and the cycle of operation begins anew. When one of the reservoirs is filling with water, the other is discharging its contents into the pipe E.

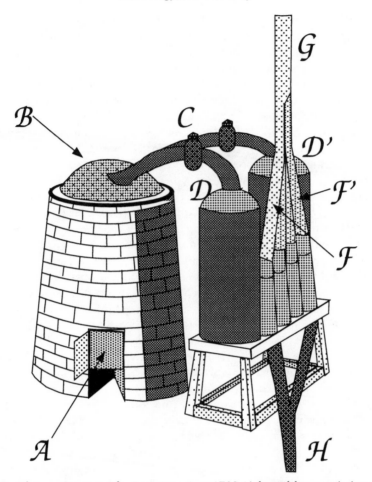

Fig. 8-3. *Thomas Savery's "fire" engine, circa 1702* (Adapted by permission of the publisher, Cornell University Press from R. H. Thurston, *A History of the Growth of the Steam Engine,* © 1939 Cornell University). Steam from the boiler enters one of the receivers (say *D*), condenses, and a vacuum is formed. Atmospheric pressure forces water up through *H* into *D*. Steam is readmitted into *D*, and forces water there up through the pipe *G*. The steam condenses and the cycle repeats. The two receivers *D, D'* alternate in their operations: While one is filling up, the other discharges.

and atmospheric pressure forces the water up through the suction pipe *H* into *D*. The cock *C* is opened again, the valve in the suction pipe *H* closes and the pressure of steam forces the water in *D* up through the pipe *G*. The cock *C* is again closed; the steam condenses and the cycle of operations begins once more. The two receivers work in parallel on alternative functions: while one of the receivers discharges water the other is filling.[35]

As was noted earlier, the similarity of Savory's to Worcester's engine is quite striking. My concern here is, however, not with the apportioning of credit between the two but simply to note what had transpired before Newcomen came on the scene.

It is useful at this point to summarize the state of the art about the time Newcomen began his investigations in, apparently, 1698. This was the year, we must remind ourselves, that Savery obtained his comprehensive patent. This summary is schematized in Figure 8-4. Here, an arrow from one entry to another indicates a direct known connection between the inventors (or inventions) cited in the entries.

Let us now recall the Newcomen engine (Figure 8-1). It should now be evident that several of its features have unmistakable antecedents going back to the marquis of Worcester (1650) on the one hand, and von Guericke (1672) on the other. More specifically, given that the Newcomen engine was most immediately preceded by Papin's and Savery's inventions, it is of great interest to determine whether Newcomen's engine incorporates components or ideas that can be traced to them. We find, in fact, that the Newcomen engine contains the following such features:

1. The piston and cylinder arrangement in Papin's scheme where (a) the pressure of the steam in the cylinder enables the piston to move up; (b) the condensation of steam produces a vacuum in the cylinder; (c) the atmospheric pressure on the piston drives the piston down. However, in the case of the Newcomen engine, the steam pressure in the cylinder equalizes the atmospheric pressure on the other side of the piston rather than directly exert the force to move the piston up.
2. Papin's concept, of 1675, of covering the piston with water to make it airtight.
3. Savery's scheme for raising steam in a separate boiler.
4. Savery's idea of raising water from the source or well by utilizing the pressure of the atmosphere on the water.

In addition, we also note that Sir Samuel Moreland had anticipated Newcomen in inventing the automatic or self-acting valve gear. The one feature of the Newcomen that did *not* appear to have any antecedent is the use of a jet of water to condense the steam inside the cylinder.

Any invention of the complexity of the Newcomen engine is, of course, vastly more than the sum of its parts. Newcomen's great creative achievement was to utilize and combine previous ideas and improve them to produce the first practicable engine based on steam and atmospheric pressure that actually came to be widely used to raise water, especially from mines.

There remains, however, the vital question: How much did Newcomen know of these antecedents?

The historians themselves seem to disagree on this matter. According to Robert Thurston, writing in 1878, Savery's engine may have been "perfectly well known" to Newcomen and Newcomen may even have visited Savery— they lived only fifteen miles apart. Thurston even refers to some accounts according to which Newcomen was actually employed by Savery for making some of the more complicated forgings for his engine; and, possibly, Newcomen and his assistant Calley paid a visit to Cornwall, where they saw a working Savery engine which directed their attention to the problem itself.[36] R. T. C. Rolt has

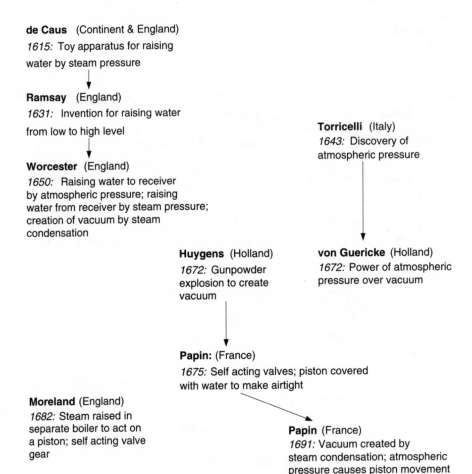

WATER RAISING **USE OF ATMOSPHERIC PRESSURE**

de Caus (Continent & England)
1615: Toy apparatus for raising water by steam pressure

Ramsay (England)
1631: Invention for raising water from low to high level

Worcester (England)
1650: Raising water to receiver by atmospheric pressure; raising water from receiver by steam pressure; creation of vacuum by steam condensation

Torricelli (Italy)
1643: Discovery of atmospheric pressure

Huygens (Holland)
1672: Gunpowder explosion to create vacuum

von Guericke (Holland)
1672: Power of atmospheric pressure over vacuum

Papin: (France)
1675: Self acting valves; piston covered with water to make airtight

Moreland (England)
1682: Steam raised in separate boiler to act on a piston; self acting valve gear

Papin (France)
1691: Vacuum created by steam condensation; atmospheric pressure causes piston movement

Savery (England)
1698: Steam raised in separate boiler; atmospheric pressure raises water to receiver; steam pressure raises water from receiver; vacuum by condensation of steam

Fig. 8-4. *Landmarks in the phylogeny of the Newcomen engine.* Two distinct pathways are visible: One, beginning with de Caus, is concerned with the problem of raising water. The other, originating with Torricelli, relates to the discovery and exploitation of atmospheric pressure.

also written of this possibility, although he hastens to add that there is no actual evidence of this, or that the two men knew each other.[37]

Furthermore, Thurston and H. W. Dickinson both quote contemporaries of Newcomen and Savery who claim that Newcomen had no knowledge of Savery's engine and that the former attempted to construct a steam engine at least as early as 1698, the year of Savery's patent.[38] Finally, we should note that an engraving of a steam engine installed near Dudley Castle in the country of Staffordshire in 1712, executed in 1719 by one Thomas Barry bears the caption "The STEAM ENGINE *near Dudley Castle Invented* by Capt. Savery and Mr. Newcomen Erected by ye *Later* 1712."

In other words, one cannot reach any definite conclusion as to whether or what extent Newcomen actually knew Savery or his work. Suffice it to say that the possibility existed.

A propos Newcomen's awareness of Papin's work, the evidence, though still inconclusive, is somewhat more positive. Both Robert Thurston and Abbott Payson Usher record that though Newcomen lived and worked in the country and was a provincial ironmonger by trade, he kept in touch with the science of the day by means of correspondence with the distinguished scientist and inventor Robert Hooke (1635–1703), a fellow of, and for several years, secretary to the Royal Society.[39] Indeed, according to Thurston, it was in the course of their correspondence that Newcomen proposed building a steam engine consisting of a cylinder and piston similar to Papin's engine that would drive a separate pump.[40] In Usher's account, Newcomen had originally conceived a different scheme and it was Hooke who, arguing against this proposal, had suggested to Newcomen the idea of operating the piston by means of a vacuum produced by the condensation of steam.[41]

This account has been flatly contradicted by Dickinson and Rolt. According to Dickinson, the connection between Newcomen and Hooke was related by John Robison (1739–1805), an eminent scientific writer, professor of natural philosophy at Edinburgh University, and friend of Newcomen's heir, James Watt. In his article on the steam engine in the third edition of the *Encyclopedia Britannica*, Robison states that Newcomen obtained from Hooke notes on Papin's work and suggestions as to how to proceed with his invention. However, both Dickinson and Rolt state that repeated searches through the minutes and records of the Royal Society had failed to produce any such documents.[42] Clearly such papers, if they had existed, may well have been lost. And it is difficult to see why Robison would concoct such a story in the first place. As Usher notes, there seems no real ground for doubting Robison's statement on this matter.[43]

There is, finally, one other way in which Newcomen could have become familiar with Papin's work; he could have seen, in the *Philosophical Transactions* of the Royal Society for the year 1697, a review of Papin's article of 1690 describing his engine.[44]

My own conclusion on the matter of whether Newcomen knew about Savery's and Papin's (or even prior) work is that while no *definite* verdict one way or the other can be reached, there is enough evidence of possible connections between Newcomen on the one hand and Savery and Papin on the other to sustain the belief that he drew upon these previous inventions in the course of his own work. Stated differently, *given* that the Newcomen engine included so many features from antecedents as I have listed above, and *given* the evidence of possible familiarity on Newcomen's part with Papin's apparatus and Savery's engine, it is difficult to harbor the belief that Newcomen's invention was *not* the outcome of evolutionary history.

ENIAC: The First General-Purpose Electronic Computer

On February 15, 1946, an electronic machine called ENIAC, an acronym for Electronic Numerical Integrator and Computer, was formally commissioned at a ceremony at the Moore School of Electrical Engineering of the University of Pennsylvania—thereby entering the pages of history as the world's first fully operational *general-purpose electronic digital computer*. And just as Newcomen's atmospheric engine was a portent of the Industrial Revolution that was soon to follow, so the invention of the ENIAC marked the then unsuspected genesis of what may be called the Age of Automation.

The ENIAC was *electronic* in that its computational capabilities were effected by the large-scale use of vacuum tubes, some 18,000 of them. It was *digital* in that it stored, transmitted, and processed discrete rather than continuous electrical entities, the former representing digital information. And it was *general-purpose* in that it had the means to solve a variety of numerical problems rather than one or a few.[45]

The principal designers of the ENIAC were John Mauchly, a physicist, and Presper Eckert, an engineer—although several other persons participated in and contributed to the ENIAC project, which "officially" spanned the years 1943–46. These included the fabled applied mathematician John von Neumann, Arthur Burks, a philosopher by training, and Herman Goldstine, a mathematician, all three of whom were to play vital roles in the subsequent development of the electronic computer. For all practical purposes, however, Mauchly and Eckert are conventionally regarded as the coinventors of the ENIAC.

The ENIAC provides a remarkable example of the role of history in technological creativity for, as we shall see, while the machine itself in its totality was, without a shadow of doubt, historically original, some of the most important ideas embodied in it had originated in a variety of antecedent artifacts. To ask, as I had in the case of the blast furnace, how and why did the first electronic general-purpose computer take the form it did is to delve into the era that preceded that of the ENIAC. This is precisely what I shall do here.

The story of the ENIAC is also distinctive because of its central position, not only in the intellectual debate about who invented the electronic computer but also in a celebrated court trial over the ENIAC patent.[46] Thus, the publicly available documentation of the history and prehistory of the ENIAC is quite extensive.[47]

For the reader to understand the evolutionary roots of the ENIAC, I will first give a brief account of its principle features. We will then consider how or in what way some of these same features had antecedents in prior computational artifacts and how these antecedents found their way into the ENIAC.

From a logical or organizational point of view, the ENIAC comprised four classes of functionally distinct entities: input-output units; a read-write memory; computing units; and a program control unit.

Data was fed into the machine from a punched-card reader and thence into a set of relays called, collectively, the "constant transmitter," (Figure 8-5) which made the data available for computation as required.

In modern terms, the constant transmitter served as an "input buffer." Conversely, the results of computations were transmitted by the "printer" (a misnomer for the ENIAC's output control device) to a card punch.

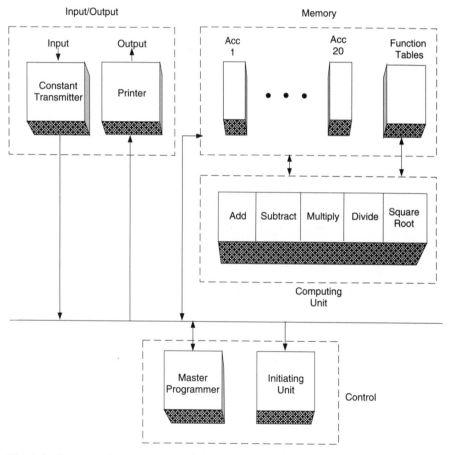

Fig. 8-5. *Functional organization of the* ENIAC. The set of twenty accumulators forms the computer's "read-write" memory. Additional "read-only" memory is provided by way of function tables. The circuits enable arithmetic operations to be performed on the accumulator contents. Input and output functions are performed by the constant transmitter and printer, respectively. The master programmer serves as the control unit.

The heart of the machine was a set of twenty "accumulators," which constituted the ENIAC's sole internal read-write memory.[48] Thus, its internal storage capacity was only twenty signed ten-digit numbers. However, additional storage capacity of up to 300 numbers was provided by way of three "function tables." These circuits, as the name suggests, were primarily intended to hold "read-only" tabular data. The function tables were set to hold values by means of manual switches.

By their very design, the accumulators had the built-in capacity to perform addition and subtraction. Other computing capabilities were provided by means of a multiplication unit and a divider and square rooter. Both these units performed their functions on the values stored in the accumulators and placed their results back into the accumulators.

The ENIAC did not have a program memory—the machine was the precursor to the stored-program computer. Each of the units had its own control comprising a set of switches that could be set manually. In order to perform a computation, the program switches associated with the relevant unit would be manually set in order to specify which operations of a unit's repertoire were to be performed. Furthermore, the various units would be connected by means of cables in order to enable transmission of signals between the units. Finally, the sequencing of these operations—that is, determining the temporal ordering of the operations—would be specified by setting switches on the "master programmer unit." This unit was, consequently, the functional equivalent of what later came to be known as the control unit in stored-program computers.

The remaining component shown in Figure 8-5 is the "initiation unit," which was responsible for clearing the machine, turning on power, and initiating a computation.

To illustrate the programming and operation of the ENIAC, let us consider a trivial example.[49] It is required to compute a table of squares. To understand how this may be effected, assume first that Accumulator 1 already contains n and Accumulator 2 initially contains n^2. The machine will then compute $(n + 1)$ and $(n + 1)^2$ as follows:

Suppose that by setting a switch, the initiation unit is made to transmit a signal causing Accumulator 1 to transmit its contents twice and Accumulator 2 to receive these contents twice. On completion of this operation, Accumulator 2 will contain $n^2 + 2n$. If Accumulator 1 now sends a signal enabling the constant transmitter to send a 1, and both accumulators to receive it, then Accumulator 1 will contain $n + 1$ and Accumulator 2, $n^2 + 2n + 1$, that is, $(n + 1)^2$.

This "program" can be used iteratively to construct a table of squares, say for $n = 0, 1, \ldots, 99$ by setting switches in the master programmer. Assume that the accumulators have initially been set to 0. The initiation unit enables the master programmer to start counting the number of outputs, to initiate the first output (that is, to record $n = 0, n^2 = 0$), and to activate the first computation. This causes the accumulators to operate as described above: Accumulator 1 transmits 2×0, Accumulator 2 receives the number, the constant transmitter sends a 1 to each accumulator, and the master programmer advances one step. Thus, $n = 1$, $n^2 = 1$ is produced. This process will continue until 100 pulses have been emitted by the master programmer, the outcome being a table of squares for the number 0, 1, . . . ,99, produced as output through the card punch.

As in the case of the Newcomen engine, if we assert that the ENIAC evolved from the earlier inventions and technological ideas, we are obliged to provide evidence that not only had certain aspects of the ENIAC been already invented and were generally in the public domain, but that one or more of the principal designers of the ENIAC were aware of these ideas. As I have already remarked, we are fortunate in this regard in that the records on the origins of the electronic computer are quite extensive. Of particular value to our objective is the detailed historical account written by Arthur Burks, one of the major participants in the development of both the ENIAC and the stored-program computer that followed, and Alice Burks, who was also a member of the ENIAC project.[50]

Of all the influences on the ENIAC—and there were many, as we shall see—three stand out, and I will deal with these first: the differential analyzer, a mechanical analog device invented in 1930; a special-purpose electronic digital computer conceived by John Atanasoff in 1937;[51] and Atanasoff's idea for modifying his special-purpose computer to solve the problems for which the differential analyzer had been designed.[52]

The *differential analyzer*, invented in 1930 by the engineer Vannevar Bush at the Massachusetts Institute of Technology,[53] was an electrically powered but otherwise purely mechanical analog computational device for solving differential equations. Its principal computing unit was a set of "integrators" for performing integration. Bush's original machine had six integrators. Subsequent versions built at the University of Pennsylvania for its own Moore School of Electrical Engineering and for the Ballistic Research Laboratory had fourteen and ten integrators, respectively. In addition, systems of gears were available for performing arithmetic operations. Functions were manually supplied to the analyzer by "input tables," and results were printed graphically on an "output table." The various units in the analyzer were connected by means of long, rotating shafts.

While the mechanical integrator itself was an ingenious device, its principles are not important to this discussion. It is sufficient for the reader to note that (a) the inputs to the integrator were the function y to be integrated and the variable of integration x, and its output was the $\int y dx$; and (b) the input and output values were supplied to the integrator in terms of linear displacements and rotational amounts of its various mechanical constituents.

The differential analyzer played two significant roles in the later invention of the ENIAC. First, the ENIAC's organization and operating procedures paralleled those of the differential analyzer; second, the developments that led to the ENIAC began by way of *dissatisfaction* with the performance of the analyzer. I shall return to both these issues later, but given the first of these two influences, it is useful to consider, very briefly, how a differential analyzer would be set up to solve a problem. Specifically, solving a particular set of differential equations on the analyzer entailed, essentially, the following major steps.[54]

1. *Problem preparation*—in which the equation would be transformed into an integral form, since the basic computational unit in the differential analyzer was an integrator.
2. *Determining the interconnection of the units*—so as to solve the equations. One may think of this as what later came to be called the program preparation step.
3. *Manually connecting the physical units*—this would entail placing gear wheels, addition units, and lengths of shafts into position and tightening them manually.
4. *Problem running*—this would necessitate setting up the initial conditions.

If the differential analyzer was one of the significant evolutionary antecedents to the ENIAC, the work of John Atanasoff and Clifford Berry (particularly the former) was another. Indeed, according to the ruling handed down by Judge Earl Larson in 1973 in the court trial between Honeywell and Sperry Rand over the ENIAC patent, the electronic digital computer was invented by Atanasoff, from whom John Mauchly and Presper Eckert derived the idea.[55] While it is obviously useful to keep this legal decision in mind, we need to see for ourselves how Atanasoff's ideas made their way into the ENIAC. For this purpose, it is necessary to establish the main ideas for which John Atanasoff is now acknowledged to be responsible, ideas which he and Clifford Berry implemented in what is now called the Atanasoff-Berry Computer (ABC). This computer was built at Iowa State University between 1939 and 1942.[56]

1. Atanasoff pioneered the use of vacuum tube circuits for digital computing. By the mid-1930s, the use of vacuum tubes, resistors, and capacitors in radios was well established. However, these circuits operated in analog or continuous mode. Digital circuit elements such as flip-flops, counters, and simple switching circuits had also been previously implemented but the digital use of vacuum tubes for building such circuits was quite rare.[57]
2. Storage memory in the Atanasoff system was in the form of electrical condensers or capacitors mounted on two rotating Bakelite drums, 8 inches in diameter, 11 inches long. A positive charge on a condenser corresponded to "0," a negative charge to "1." Each drum could store thirty 50-bit numbers. As the drums rotated, these 50-bit numbers could be read, processed, and replaced serially. The two drums were mounted on the same axle so that they could operate synchronously.
3. The ABC was a special-purpose computer for solving systems of simultaneous linear equations by the method known as Gaussian elimination. For this purpose, the coefficients of two equations from the set were read from cards onto the two drums, one of which was named the "keyboard abacus" (KA), the other, the "counter abacus" (CA). To understand the arithmetical process at work, consider just a pair of equations:

$$a_{11}x_1 + a_{12}x_2 = b_1$$
$$a_{21}x_1 + a_{22}x_2 = b_2$$

Assume that x_1 was to be eliminated, that a_{11}, a_{21} were represented by the leftmost numbers on the drums KA and CA respectively, and that both a_{11}, a_{21} were positive. The aim, then, was to reduce the coefficient a_{21} on CA to 0. This would be achieved by first *subtracting* the KA coefficients from the CA coefficients repeatedly until the latter became negative. The KA numbers would then be *shifted* one binary position to the right and *added* to the CA numbers again until a change in sign occurred in the leftmost coefficients. Another right shift would then be done. This process continued until the leftmost coefficient was eliminated from CA. The resultant equation on the CA was punched onto a card and the process repeated for another pair of equations.

4. In order to perform the operations just described, the computational capabilities in the ABC was comprised of thirty add-subtract units that could add (or subtract) numbers on the KA to (or from) those in the CA in parallel. Circuits were also available for shifting numbers on the KA.

5. Input data were fed to the machine through punched cards in decimal form. The ABC was itself a binary machine, hence decimal-to-binary conversion was done before a number was stored on the drum.[58] Similarly, the results of computations, stored on the drums, were converted to decimal form before being punched as output on cards. Intermediate results produced in the course of a computation were, however, stored in binary form.

As Arthur and Alice Burks have pointed out:

> Atanasoff had achieved a great deal in his pioneering efforts. He invented a novel form of serial store suitable for electronic computation. He conceived, developed and proved the fundamental principles of electronic switching for digital computation, principles encompassing arithmetic operations, control, base conversion, the transfer and restoration of data, and synchronization. He designed a well-balanced electronic computing machine utilizing this store and these principles and having a centralized architecture of store, arithmetic unit, control circuits, and input-output with base conversion.[59]

But there was still more to Atanasoff's creative process. In the spring of 1941, while in the midst of building the ABC with Clifford Berry, he became aware, through discussions with the engineer Samuel Caldwell, of the differential analyzer, an improved electromechanical version of which was then being built at the Massachusetts Institute of Technology by Vannevar Bush and Caldwell. Atanasoff conceived the idea of using the ABC to solve the differential equations that were being solved by the analyzer. This would necessitate the application of numerical methods of solving differential equations, methods with which he was already familiar and which at that time were usually conducted on mechanical desk calculators. In such numerical methods, the integral

$$\int_{x=a}^{x=b} f(x)dx$$

is replaced by the summation

$$\sum_{i=1}^{k} f(x_i)\Delta x$$

where the interval of integration a to b is divided into k equal subintervals of length Δx, designated as $x_1, x_2, \ldots, x_k$.

Atanasoff realized that such numerical integration could be effected by using a combination of accumulation, multiplication, and addition.[60]

This, then, was the situation on June 13, 1941, when John Mauchly visited Atanasoff to learn about the latter's computer. The two had previously become acquainted at a scientific meeting in December 1940, when Mauchly had come to know of the work being done on the ABC. Mauchly was even aware prior to his visit, by way of a letter written by Atanasoff to him on May 31, 1941, of the latter's idea for using the ABC to solve differential equations.[61]

Mauchly stayed with Atanasoff and his family for four days, during which the visitor had the opportunity to examine the ABC and have extensive discussions with his host about computers.[62] He was even given the opportunity to read and make notes from a thirty-five-page report that Atanasoff had written in August 1940 on the design principles of his computer.[63] Mauchly, in other words, became quite knowledgeable about Atanasoff's invention by the time he left Iowa. In September 1941, he was to write to Atanasoff of his own ideas concerning the design of an electronic computer—ideas that he described as "hybrid," combining some of Atanasoff's ideas with others that were his own. He then asked Atanasoff whether the latter would object to his building a computer incorporating some of the features of the ABC.[64] He was also, Mauchly wrote, studying the differential analyzer then in operation at the Moore School (where he was then working) and had had the notion of outperforming the analyzer through electronic means.

There is sufficient evidence, then, that Mauchly had considerable knowledge about both the differential analyzer and the Atanasoff machine when he began to think seriously about building an electronic computer sometime in mid-1941. In August 1942, he wrote a memorandum titled "The Use of High Speed Vacuum Tube Devices for Calculating," which contained a detailed discussion of the advantages of using electronic circuitry for computing devices.[65]

As to the beginnings of the actual project that led to the ENIAC, Arthur Burks has documented how it arose:

> After we entered World War II, the Moore School became a center for calculating firing tables. The differential analyzer was soon used full time for trajectories. . . . Herman Goldstine was the Army Lieutenant in charge of this activity.
>
> One day John [Mauchly] suggested to Herman that trajectories could be calculated much faster with vacuum tubes. Herman thought John's suggestion a good one, and asked for a proposal. John, Pres Eckert and J. G. Brainerd then wrote Report on an Electronic Diff* [sic] Analyzer (2 April 1943). This

was submitted to the Ballistic Research Laboratory, Aberdeen Proving Ground, on behalf of the Moore School of Electrical Engineering, University of Pennsylvania.

In this proposal the Moore School offered a machine that could compute ballistic trajectories at least 10 times as fast as the differential analyzer ... Herman persuaded the Ballistic Research Laboratory to fund the proposal. The electronic diff* analyzer later became the Electronic Numerical Integrator and Computer, or ENIAC for short.[66]

We thus see that the ENIAC had its genesis in dissatisfaction, on Mauchly's part, with the differential analyzer. Technological creativity, as noted in Chapter 3, so often begins with the urge to improve on the way things are.

As I remarked earlier in this discussion, the Atanasoff machine and the differential analyzer were, in a global sense, the two most profound ideational sources for the development of the ENIAC. But they were by no means the sole influences. Let me then summarize the most significant of the evolutionary traces of the past in the ENIAC, not only those derived from Atanasoff and Vannevar Bush, but others also.

The Use of Vacuum Tubes

The vacuum tube, the basic digital element used to construct flip-flops and switches in the ENIAC, had been previously used (for the first time in digital computing) in the ABC as a digital element for its arithmetic and control circuits. And, as we have seen, Mauchly was fully aware of Atanasoff's work in this regard. In the ENIAC, vacuum tubes were employed not only for the switching circuits but also to build the read-write memory: each of its twenty 10-digit accumulators was composed of ten decimal electronic counters (and a binary counter for the sign), the basic element of which was the flipflop built with vacuum tubes.

The Adoption of Numerical Integration

The idea of solving differential equations using digital arithmetic—and thereby improving on the speed and efficiency of the differential analyzer—upon which the arithmetical or computing capabilities of the ENIAC were founded, can also be traced back to Atanasoff and to the fact that the latter had communicated the idea to Mauchly.

The Combining of Storage and Arithmetic Functions

As we have seen, the central unit in the ENIAC was the accumulator. This was a component that combined both storage and arithmetical functions: two accumulators in cooperation, for instance, constituted an adder-subtractor. This combination of storage and computation was also present in the differential analyzer: an integrator both computed something and stored the resulting value in the form of linear and rotational positions. Similarly, fixed and differential gears computed and stored inputs and outputs as rotational positions.

The Adoption of Decimal Computing

From the perspective of the logical design, or, as it would now be called, architecture of the ENIAC, Mauchly admitted in his 1942 memorandum that he conceived his electronic computer by directly drawing an analogy with the mechanical counterparts then in use. In particular, he adopted the decimal system in use in mechanical computers. Thus, for example, the number '465' would be stored in an accumulator by a count of five in the units register, six in the tens register and four in the hundreds register.[67] Here we see, then, the origins of the decimal nature of the ENIAC and, in particular, the adoption of decimal accumulators.

The Use of Punched-Card Input/Output

The ENIAC input-and-output mechanism originated in the electromechanical technology widely used in the IBM data-processing machines and the relay circuits developed by 1940 at the Bell Telephone Laboratories. In the case of the IBM machines, holes in cards were sensed electrically, and the resulting signals were transmitted over wires to operate a variety of devices. Electrical signals also drove punches that made holes in cards. And at the Bell Laboratories, relays were used to build a series of computing devices from about 1937.[68] One of these machines, the "complex computer," began operation in January 1940 and was publicly demonstrated in September 1940 in New York on the occasion of a meeting of the American Mathematical Society. John Mauchly was present at the demonstration.[69]

In the case of the ENIAC, data was fed into the computer from punched cards read by an IBM card reader. These numbers were transformed electrically into relays in the constant transmitter. The converse process occurred in the case of producing output—results were generated by way of an IBM card punch.

Origins of the ENIAC Multiplier Unit

The multiplier unit in the ENIAC was the most complicated unit in the machine in terms of the amount of circuitry that it contained. It was, in fact, an electronic version of a fast multiplication algorithm invented in 1889 by Léon Bollée—an algorithm, as Arthur and Alice Burks pointed out, that was used in calculating devices ranging from mechanical, through electromechanical, to the electronic versions, including the ENIAC.[70] In particular, there were strong similarities between the way multiplication was effected in the ENIAC and the operation of one of the most widely used electromechanical data-processing machines of the 1930s. To appreciate this, it is necessary to consider the nature of the algorithm itself.[71]

Suppose that it is required to multiply the number 42 (the *multiplicand*) by the number 56 (the *multiplier*). Let us further suppose that the conventional multiplication tables for multiplying two digits are available. Given the product of two digits, its lower-order digit can be designated as the "right hand"

(RH) component and the higher-order digit the "left-hand" (LH) component. For example, in the product 5 × 3, the RH component is 5 and the LH component is 1.

Bollée's algorithm for multiplying 42 by 56 would proceed as shown schematically in Figure 8-6.

Thus, the partial products for the simultaneous multiplication of the multiplicand first by the units digit of the multiplier and then by its tens digit are accumulated as the RH and LH components; these components are then added. The reader will note that the procedure utilizes the multiplication tables to compute the partial products.

The ENIAC's multiplier used four accumulators (or six, depending on whether a full 20-digit product was required); one each for the multiplicand and multiplier, one (or two) for the LH component, and one (or two) for the RH component. The multiplier itself consisted of subunits for sensing the multiplier and multiplicand, a circuit that stored the single-digit multiplication tables, and shifters for shifting the partial products. Arthur and Alice Burks have pointed out that the same method of multiplication was employed in the IBM 601 electromechanical "crosspoint" multiplier unit, which, according to them, was "not only . . . the most powerful data processor of the 1930s but . . . the most powerful calculator on the market."[72] In the case of the 601, its "plugboards" were prewired to perform specific calculations as would be

multiplicand: 42		*Accumulation of partial products*	
multiplier × 56			
		LH	RH
partial product for multiplier units	6 × 2 = 12		42
digit 6 done simultaneously	6 × 4 = 24–	210	
partial product for multiplier tens	5 × 2 = 10–		00–
digit 5 done simultaneously	5 × 4 = 20– –	2100	
		2310	042
	=	2352	

Fig. 8-6. *An application of Bollée's multiplication algorithm.*

required for such typical data-processing applications as payroll calculations. For example, a plugboard could be wired to calculate the expression $A \times B + C$. The input values for A, B, C would be read from punched cards, the multiplication performed using Bollée's algorithm, the result placed in LH and RH partial product "counters" (the functional equivalent of ENIAC accumulators), the addition of partial products performed and the result punched onto cards. The plugboard itself had multiple rows of sockets some of which received digits from input cards, others sending digits to the punch unit while the remaining ones would be connected to various computing and control circuits. There was, as the Burkses noted, significant similarity between the operation and timing sequence of the crosspoint multiplier and those of the ENIAC multiplication unit.

Arthur Burks has further recorded: "Some years ago, I wrote to John McPherson, a leading IBM engineer in the punched-card era about the relation of the ENIAC's high speed multiplier to the antecedent IBM machines. He gave me information about the IBM 601 and 600, then observed, "I felt you were doing electronically almost exactly what we had had in general use for 12 years in the 600 and 601.""[73]

Eckert and Mauchly were certainly aware of the Bollée algorithm. They had originally intended to implement multiplication on the ENIAC by the "slow algorithm," that is, by repeated additions. They decided to adopt the "fast algorithm" only after learning about a circuit called the "register-transistor matrix," invented simultaneously by Perry Crawford at the Massachusetts Institute of Technology and Jan Rachman at the RCA Laboratories in Princeton. This matrix was used to hold the multiplication tables in the ENIAC.[74] As to whether the designers of the ENIAC were aware of the nature of the IBM electromechanical calculators, both Arthur Burks and Herman Goldstine, two of the participants in the ENIAC project, imply that they were. Burks, for example, wrote that "the arithmetic design of ENIAC was influenced mainly by two kinds of calculators: mechanical desk calculators . . . and electromechanical card operated IBM machines,"[75] while Goldstine remarked that John Mauchly "alone of the staff of the Moore School knew a lot about the design of the standard electromechanical IBM machines of the period and was able to suggest to the engineers how to handle various design problems by analogy to methods used by IBM."[76]

Adoption of Plugboard-based Problem Setup Technique

The IBM plugboard concept was also used by the ENIAC inventors in their design of the method of programming—or what was then called "problem setup." Once a problem had been transformed into a form suitable for the machine—differential equations, for example, transformed into difference equations—the next step was to establish how the units should be interconnected and their switches set so as to perform the desired computation. Problem setups were represented in a variety of ways, including what Burks called "setup diagrams."

The ENIAC problem setup procedure was, in fact, derived from and was an extension of, the plugboard.[77]

Origins of the Architecture of the ENIAC

We have already noted that John Mauchly's original intention had been to outperform the differential analyzer "electronically." This intention was stated in his letter of September 30, 1941, to Atanasoff.[78] Mauchly's memorandum of August 1942 entitled "The Use of High Speed Vacuum Tube Devices for Calculating" reiterated this intention; indeed, the first half of this article is essentially a catalog of the advantages of an electronic calculator over the mechanical differential analyzer.[79] The ENIAC project was officially born from the conviction that an electronic computer would perform the calculations then being done on the analyzer ten times faster. Finally, one must also take note that the proposal submitted in April 1943 by Mauchly, Eckert, and Brainerd to the Ballistic Research Laboratory bore the title "Report on an Electronic Diff* Analyzer." Here the abbreviated term "Diff*" apparently represented "difference" rather than "differential," reflecting the discrete nature of the electronic analyzer.[80]

Given this very definite and ubiquitous presence of the differential analyzer in the thinking that led to the ENIAC, it should not surprise us to find striking similarities in the overall organization or "architecture" of the ENIAC with that of its analog predecessor. This similarity is clearly remarked upon by Arthur and Alice Burks, who note that

> the 1943 proposal itself drew an analogy between their [that is, the differential analyzer's and the proposed machine's] architectures . . . in a one-page section titled "Brief Description of Electronic Analyzer." There Major Albert A. Bennett, a mathematician serving at Aberdeen during the war, was cited for his description of the differential analyzer as a group of interconnected submachines; and then it was observed that along these lines, the electronic difference analyzer could be described as a group of interconnected circuits. The proposal did go on to comment that this type of description was reasonably exact but hardly served as more than an introduction. Yet from an architectural point of view, the analogy between the mechanical differential analyzer and the ENIAC held.[81]

The architectural similarity between the two can be observed, for instance, in terms of the correspondence between components of the analyzer on the one hand and those of the ENIAC on the other: Instead of the input tables of the analyzer, the ENIAC had the function tables and constant transmitter; the ENIAC's printer corresponded to the analyzer's output table and printer; the analyzer's integrator was replaced in the ENIAC by its accumulator and multiplier; the accumulator also performed the addition-subtraction functions performed in the analyzer by its system of gears. There was even an analogy between the scheme for interconnecting the shafts in the analyzer and the method of interconnecting the numerical components in the ENIAC.[82]

Figure 8-7 summarizes, schematically, the phylogenic pathways that led to the invention of the ENIAC.

Phylogeny Conditions Ontogeny

If one accepts that the two historical accounts just presented, one of the origins of the atmospheric engine and the other of the birth of the ENIAC, exhibit the phenomenon of phylogenic evolution—that, in fact, in each case there is a linked network of mature artifacts or artifactual forms leading to the invention in question—one must also seriously entertain the larger conclusion which, for convenience, I shall call

The phylogeny law: Every act of invention or design has a phylogenic history.

And since, drawing an analogy with the biological world, I have previously called the process whereby an artifactual design comes into being "ontogenic evolution"[83] another statement of this law is that *phylogeny conditions ontogeny.*[84]

This does not mean that a cognitive agent embarking on a technological problem is necessarily explicitly aware of all that has happened in the relevant past. But it does imply that the knowledge that he or she brings to bear on the problem at hand will include tokens that represent, or embody, or are the most recent elements of, a phylogenic past. And that without that past, these tokens would not be present in the agent's knowledge body. And that it is these tokens that condition or influence the agent's process of ideation.

Thus, for Maurice Wilkes, his invention of microprogramming was conditioned by his knowledge of the EDSAC and Whirlwind computers. But these, in turn, were the elements of a phylogenic past that included the Atanasoff-Berry Computer and the ENIAC. Robert Stephenson's tubular bridge emerged from the suspension and truss bridge forms and the evolutionary past that these belonged to.[85] Benjamin Huntsman's invention of the process for making cast steel drew upon the previously invented technology of brass founding. Alexander Graham Bell's development of the telephone was conditioned by Helmholtz's apparatus for producing vowel sounds. The Mond Company metallurgists began the process that led to the first superalloy by utilizing their knowledge of previously developed nickel- and chromium-based alloys. And in this chapter, we have examined two examples of how such "immediate" knowledge that the agent brings to bear in a given situation is itself the product of phylogenic evolution.

Testability and Other Aspects of the Phylogeny Law

Several points are worth noting concerning the phylogeny law. First, like the hypothesis law, this law is intended to be an empirical and therefore testable proposition about the design and invention of artifacts. As in the case of the

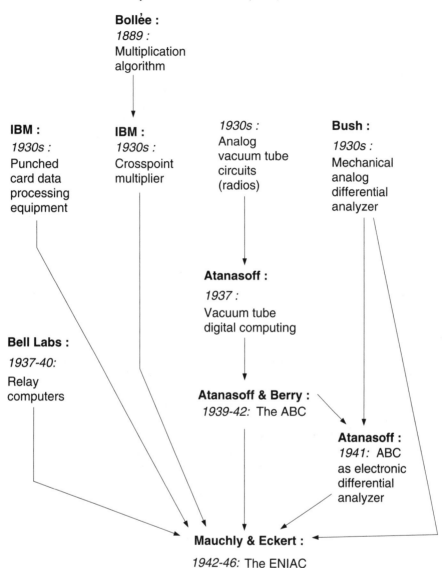

Fig. 8-7. *Landmarks in the phylogeny of the* ENIAC. Several pathways are visible. Perhaps the most significant is the one beginning with analog vacuum-tube circuits since this determined the use of vacuum tubes for digital computation in the ENIAC. The next most significant pathway is probably the one on the right, showing the influence of the organization and operation of the differential analyzer on the ENIAC.

hypothesis law, its testability hinges upon specific case studies of artifacts: The law will be corroborated by evidence of phylogenic evolution; conversely, any particular instance of an invention that cannot be ascribed to prior artifactual forms will constitute a failure to corroborate the law, thereby casting doubt on its universal validity.

Second, the phylogeny law may be said to simply cast in testable form something that others, particularly historians of technology, had come to recognize for some time—namely, the significance and role of evolution in technological change. I have already mentioned George Basalla's book *The Evolution of Technology*, which attempts to explain technological change in explicitly evolutionary terms.[86] A more specific study is Edward Constant's examination of the circumstances leading to the birth of the turbojet engine in the 1930s by the Englishman Frank Whittle and the Germans Hans von Ohain, Herbert Wagner, and Helmut Schelp.[87] The advent of the turbojet heralded a technological revolution in Thomas Kuhn's sense.[88] But this revolution, in Constant's words, "depended upon prior technology, upon a two-centuries-long evolution of turbine systems: water turbines, turbine pumps, steam turbines, turbo-air compressors, internal combustion gas turbines and turbo supercharges. It required the prior successful co-evolution of piston aero-engines and steamlined [*sic*] airframes."[89] And Henry Petroski has devoted two books to the examination of the role of evolution in the development of the more mundane artifacts of everyday life such as paper clips and table forks,[90] and the pencil.[91]

Third, the phylogeny law, if true, drives yet another nail into the coffin of the ineffability theory of creativity, at least in the technological domain. For, we now have yet another explanatory dimension for the creative act: the evolutionary past. To be sure, we cannot explain creativity *only* in terms of the past—one must still need to appeal to other kinds of knowledge and other assumptions about the nature of the cognitive act of invention, as we have seen in the preceding chapters. But much of how or why artifacts are given the form they are, is rendered explicable once the fact of evolution in the phylogenic sense is recognized.

Fourth, although it is being claimed that phylogeny *conditions* ontogeny, there is nothing inevitable or deterministic in the *way* that the evolutionary past may influence the technological present. For a given invention, the historical time that preceded it and the geographical space in which it is placed are populated by a variety of related artifacts, invented or designed, and ideas conceived. Some of these may indeed link to form evolutionary pathways. But perhaps only one formed the phylogenic network that actually led to the artifact of interest.

The origins of the ENIAC, the first electronic general-purpose computer, illustrates this contingency aspect of the phylogeny law rather well. Anyone acquainted with the history and prehistory of the computer will be aware of Charles Babbage, the remarkable English mathematician and "scientific mechanician" who conceived and designed first the "difference engine" and then the "analytical engine" between 1822 and 1842. Both these were *mechanical digital computers*, the first being a special-purpose device; the latter, a program-controlled general-purpose computer.[92] As Brian Randell's invaluable anthology indicates, Babbage's work formed the starting point for a number of investigations and inventions pertaining to computers along the lines of the analytical engine from 1871, the year Babbage died, to as late as 1938.[93] And Maurice Wilkes, a later computer pioneer who also studied Babbage's notebooks in some detail, has noted that Babbage was clearly exploring the logical and organizational aspects of computer design—issues that would reemerge well over a century later.[94] Yet this particular evolutionary pathway, or at least Babbage's part in it, appeared to have only influenced the development of electromechanical calculators by Howard Aiken at Harvard University in his cooperative venture with IBM.[95] It did not appear to have any influence on the ENIAC. As far as the phylogeny of the ENIAC is concerned, the beginnings of this, as we have seen, happened to be a mechanical *analog* machine—the differential analyzer. As my account earlier in this chapter indicates, there is no evidence that Babbage's work — or for that matter Aiken's invention—had had any presence in the developments that led to the ENIAC.[96] But of course Babbage, either directly or through others who were influenced by him, such as Aiken, could easily have had an influence under different circumstances.

As regards John Atanasoff, he has recorded in a relatively recent retrospective account that at the time he was thinking about his computer, he had "a cursory knowledge" of Babbage's work.[97] There is no indication, though, that such knowledge was brought to bear on his own thinking.

Finally, it must be reiterated that the hypothesis that an act of creation is in part explainable in terms of what happened in history in no way undermines the significance of the creative act. Newcomen's engine as a complete artifact far surpassed what had been achieved before, and its historical originality was and remains a matter beyond dispute. Likewise, the ENIAC as a complete artifact far exceeded the Atanasoff-Berry Computer, the differential analyzer, and the plugboard machines in its generality and speed of operation. It was, after all, the *first* general-purpose electronic computer. Mauchly and Eckert's place in history is assured. The processes of ideation that led to the atmospheric engine, to the ENIAC, to the microprogrammed control unit, to the turbojet engine, to the first superalloy, to the Britannia Bridge—all were enacted in the minds of individual cognitive agents and not in some collective historical consciousness.

9

The Nature of
Technological Knowledge

Suppose we are awarded a research grant to develop a computer program that is to be creative in some specific technological domain. We are about to embark, in other words, on the design of a "machine that will invent." An agenda such as this for mechanizing creativity is not far-fetched in the contemporary scientific milieu; there are, after all, many workers in the field of artificial intelligence (AI) who have been pursuing precisely such goals.[1]

Like any other research endeavor, our approach to the problem at hand is likely to be grounded in a particular scientific paradigm.[2] Let us suppose that we subscribe seriously (as indeed I do) to the *knowledge-level hypothesis on ideation* (KLHI) described in Chapter 4, and that KLHI and its enveloping concept of the knowledge-level model of cognition constitute the paradigm within which our inquiry will proceed.

The purpose of this scenario is not, of course, to actually build such a machine but to provide a practical context to the question I wish to address in this chapter. *What kinds of knowledge does the technologist bring to bear on the act of creation?* For surely, if we were to attempt to design a machine that invents within the knowledge-level paradigm, such a machine must also have the capability to acquire and use the same kinds of knowledge as the creative technologist.

So, to repeat, what does the creative technologist know? We already have (in Chapter 4) a partial answer to this question. The knowledge the technologist possesses includes facts, theories, hypotheses, laws, rules, paradigms, and worldviews. But the issues at stake here are somewhat more problematic: What *kinds* of facts, theories, hypotheses, rules, and so on form the corpus of technological knowledge? Can we outline, in some sense, a taxonomy of such knowledge? What is it that distinguishes what the technologist knows from what the scientist knows? In terms of knowledge, is the modern inventor and designer a mere applier of scientific knowledge?

The reader may realize that I have come almost full circle by returning to some of the questions posed in the Prolegomenon. But now, I wish not to abstractly speculate on the answers but to treat these as matters of empirical inquiry. Perhaps the case studies populating the preceding chapters, together with other empirical studies conducted by historians of technology, will shed some light on these matters.

Scientific Knowledge in Technological Reasoning: The Reductionist View

Engineering, according to the computer scientist Mary Shaw, "relies on codifying scientific knowledge about a technological problem domain in a form that is directly useful to the practitioner, thereby providing answers for questions that commonly occur in practice."[3]

Here we have a version of what is perhaps the most widely held view of the relationship between science and technology—between scientific *knowledge* and technological *practice*. In general terms, this view is that the various technological disciplines, taught at the universities and engineering schools, and practiced professionally, entail the harnessing or application of natural science to the design and construction of artifacts. According to this view, then, electrical engineering is the application of electrophysics to the design and construction of electrical circuits, machines, and systems; mechanical engineering is the application of classical mechanics and thermodynamics to the design of moving machines and mechanisms; and so on (see also Chapter 1). I shall call this view the *reductionist* theory of technological knowledge, for it asserts that the knowledge that the technologist brings to bear on the problem is, or reduces to, the knowledge generated in the natural sciences. The technologist receives and merely applies this knowledge.

That the reductionist view is widely held by *natural scientists* is, from an ideological perspective, hardly surprising. Edwin Layton has recorded how reductionism appeared on the American scene early in the nineteenth century through the writings of Joseph Henry, one of the founders of the scientific profession in the United States. According to Henry, "every mechanic art is based upon some principle of one general law of nature."[4] But what *is* unexpected is that many technologists, including those who have exerted influence on the shape of "official" policies on science and technology, also subscribe to the reductionist theory. For example, Layton quotes Thomas C. Clarke, who in his presidential address to the American Society of Civil Engineers in 1896 asserted that "science is the discovery and classification of the laws of nature. Engineering . . . is the practical application of such discovered laws."[5] In more recent times, one of the clearest articulations of the reductionist view is due to the engineer Vannevar Bush. During the Second World War, Bush was a highly influential government advisor on scientific research and development and in this capacity produced a 1945 report titled *Science: The Endless Frontier: A Report to the President*. In this report, Bush wrote of basic research—that is, research in the basic or natural sciences—producing new knowledge; the "scientific capital" from which practical applications derive. Artifacts—products and processes—are arrived at on the basis of new principles and concepts that are the fruits of researches in the "pure" sciences.[6]

The reductionist view, as a sweeping theory of technological epistemology, leads to the following obvious conclusion: If it is the case that technology itself generates no new knowledge but is founded on scientific knowledge—that is, the knowledge yielded by natural science—and given that most of science itself is essentially the product of only the past three centuries, then we are led to conclude, as Edwin Layton noted in 1974, that all artifactual creations prior to this period—that is, almost all of technological history from the emergence of hominids—*entailed no new knowledge at all.*[7] The absurdity of this conclusion demands the rejection of the reductionist hypothesis itself. It demands that we recognize that *technological knowledge is far more than just scientific knowledge; that science is but one component of the epistemology of technology.*

In fact, even in those instances of technological reasoning that are heavily grounded in the natural sciences—in those special cases, in other words, to which the reductionist view might seem to apply—we still find the presence of distinctly other kinds of knowledge; furthermore, we will observe that the interplay of scientific and "nonscientific" knowledge is remarkably complex and subtle, to the extent that even in such instances, we can only refer to the reductionist interpretation as *naive reductionism.*

Technological Knowledge:
Basic Science and Technological Theory

Consider, once more, the invention by the Mond metallurgists of the first superalloys for gas-turbine blades. We saw in Chapter 6 the nature of the reasoning that subscribed to this particular invention. Based on that discussion, it is possible to identify the following knowledge tokens that took part in the reasoning process:

$k1$ 80/20 nickel-chromium alloys are suitable for making electrical heating alloys.

$k2$ 80/20 nickel-chromium alloys are oxidation (corrosion) resistant.

$k3$ Nickel-chromium-iron alloys are creep resistant.

$k4$ Nickel-chromium-iron alloys are age-hardenable.

$k5$ The age hardening of copper-nickel-tin alloys is a possible cause of these alloys' creep strength.

$k6$ Inconsistency in the creep strength of different batches of 80/20 nickel-chromium alloys is attributable to variations in their titanium and carbon contents.

$k7$ An iron base phase called "sigma" is highly brittle.

$k8$ Characteristic X-ray patterns can reveal the presence of Ni_3Ti and Ni_3Al phases.

$k9$ Adjustments of the titanium/aluminum ratio controls the extent of the γ' precipitate phase in nickel-chromium alloys.

$k10$ The elements chromium, cobalt, and iron control the matrix properties of nickel-chromium alloys.

*k*11   In nickel-chromium alloys, the matrix and the γ' precipitate phases have the same crystal structure.

*k*12   In nickel-chromium alloys, the γ' phase is coherent to the matrix.

*k*13   X-ray diffraction patterns reveal the presence of distinct phases in solids.

*k*14   Metals and alloys have certain characteristic crystal structures.

*k*15   Alloys are solutions in the solid state.

*k*16 Ni_3Ti and Ni_3Al are intermetallic compounds.

*k*17   The lattice parameter is a spatial property of the lattice structure of crystals.

*k*18   Titanium is a deoxidizer.

*k*19   If 80/20 nickel-chromium and/or nickel-chromium-iron alloys are age-hardened, they would be creep resistant, high-temperature-rupture resistant, corrosion resistant, and shock resistant.

*k*20   Age hardening is a cause of creep resistance.

*k*21   The Ni_3Ti and Ni_3Al phases affect age hardening.

*k*22   The γ' phase containing dissolved titanium is most effective for age hardening.

*k*23   A nickel-chromium alloy with pure nickel-titanium hardener containing some aluminum is creep resistant, high-temperature-rupture resistant, corrosion resistant, and shock resistant.

Of these tokens, *k*1 through *k*18 were *used* explicitly or implicitly by the Mond metallurgists, while tokens *k*19 through *k*23 were *generated* as hypotheses in the course of their reasoning. One might then refer to *k*1 through *k*18 as "input" tokens to the invention process, *k*19 through *k*22 as "intermediate output" tokens, and *k*23 as the "final output" token of the process.

I have selected the example of the creep-resistant alloy first, because this, as much as any other conceivable technological example, seems to lend credibility to the reductionist theory. At first glance, all of the input tokens seem to be instances of physicochemical knowledge. In that case, it would be valid to claim that the invention of the first superalloys was indeed founded on scientific knowledge. Thus, the reductionist theory appears to be legitimized in this instance at least.

The circumstances are, in fact, somewhat more subtle than the above would suggest. To understand, this, let us first note that an *alloy*, formally speaking, is a solid solution of two or more kinds of atoms in which the solvent—the more abundant atomic form—is a metal. This is a definition that the physicist or the physical chemist would find acceptable since it draws directly upon basic physicochemical concepts such as "solid solution" and "atom." However, physicists and chemists are *primarily* interested in *natural phenomena*. Though historians tell us that between 4000 and 3000 B.C. the very first alloys—copper containing arsenic, and low-tin bronzes—were in all likelihood discovered *accidentally* in the course of other metallurgical operations,[8] thereby constituting natural occurrences, thereafter, alloys have been and continue to be *deliberately created* by man. Furthermore, it is because of their practical usefulness that the understanding and the development of alloys have progressed over the past five thousand or more years. Thus, alloys must be legitimately regarded as *artifacts* and not naturally occurring phenomena.

The implication of this is that it is not necessarily the case that all knowledge tokens pertaining to alloys, even physicochemical knowledge tokens, are of intrinsic interest to physicists or physical chemists because of the light they may shed on the nature of the natural world. Nor is it the case that all such knowledge tokens were generated because of their significance to our understanding of nature. Rather, the significance of much of knowledge pertaining to alloys lie in the relevance to the practical usefulness of alloys. That is, they constitute *metallurgical* knowledge or, still more broadly, since metallurgy is a technology, *technological* knowledge.

This observation echoes the aeronautical engineer and historian Walter Vincenti's distinction between scientific and engineering knowledge: "The distinction between engineering and science becomes less objective at the level of knowledge generation . . . engineering knowledge generating activities are pursued *primarily* . . . to produce practically useful knowledge; . . . scientific activities are carried out *primarily* . . . for the sake of understanding or . . . for the sake of knowledge itself. . . . The epistemological distinction is one of priority and degree of purpose rather than method." [italics in the original][9]

Elsewhere, referring to this "epistemological distinction," Vincenti notes that the criterion of validity for engineering knowledge is whether it helps in the design, manufacture, or operation of something—an artifact—that solves some particular practical problem.[10] In contrast, the criterion for whether something constitutes scientific knowledge is whether it advances one's understanding of some feature of the natural world.

If my argument and that of Vincenti's before me are valid, what can we say about the tokens $k1$ through $k23$ listed above?

We observe first that $k13$ through $k18$ are assertions of a very general nature, the scope of which transcends "mere" artifacts such as alloys. The assertion that X-ray diffraction patterns reveal the presence of certain phases in solids, that metals and alloys have certain characteristic crystal structures, that Ni_3Ti and Ni_3Al belong to the class of intermetallic compounds, that titanium is a deoxidizer, that the lattice parameter is a spatial property of the lattice structure of crystals—all are tokens of intrinsic interest to physicists and chemists regardless of their relevance to the design and manufacture of alloys. These knowledge tokens are precisely of the kind that Joseph Henry and Thomas Clarke in the last century, and Vannevar Bush in this, referred to as "laws of nature" or knowledge obtained from the "purest realm of science." I shall, therefore, refer to such knowledge as *basic science* or *basic knowledge*. They emanate directly from the physicist's and the chemist's study of the nature of matter.

Tokens $k1$ through $k12$, and $k19$ through $k23$, in contrast, are primarily relevant to the invention, design, and production of alloys. They constitute what I referred to earlier as metallurgical or, more generally, technological knowledge. However, we observe that some of these tokens express general relationships between properties or characteristics of particular alloys or classes of alloys, for example k5   The age-hardening of copper-nickel-tin alloys is a possible cause of these alloys' creep strength.

Others describe general properties of particular classes of alloys as, for instance

k2   80/20 nickel-chromium alloys are oxidation (corrosion) resistant.

k11   In nickel-chromium alloys, the matrix and γ' precipitate phases have the same crystal structure and nearly same lattice parameters.

or explain why or how certain phenomena in alloys occur, as exemplified by the token

k21 Ni_3Ti and Ni_3Al phases affect age hardening.

It will be noticed that these tokens are indistinguishable in form or nature from the tokens of basic science identified earlier or, for that matter, from knowledge about natural entities such as organisms, minerals, or galaxies. The distinction lies in that the entities here happen to be artifacts. They are propositions that have both explanatory and predictive capabilities; and many of these tokens are even expressed in terms of the fundamental concepts of physics and chemistry.

Such metallurgical knowledge tokens, then, resembles basic scientific knowledge. However, the scope of applicability of such knowledge is restricted to the artifactual domain of alloys. Furthermore, such knowledge was generated primarily because of its relevance to the metallurgist *seeking to understand metallurgical phenomena in order to produce alloys of the described characteristics.*

I shall, accordingly, refer to this kind of technological knowledge as *technological theory.* Where the context is clear, I shall simply call such knowledge "theory" or "theoretical knowledge." In the specific context of alloys, such knowledge becomes "metallurgical theory"; in the context of bridges and other structures, this knowledge is "structural theory."

It is, of course, important to keep in mind that there is no absolutely sharp line demarcating basic science from technological theory.[11] Moreover, just as basic science is constituted of a hierarchy ranging from specific facts through laws to theories and onto paradigms, so also technological theory forms a hierarchy ranging over several levels of abstraction. In the case of the invention of superalloys, for instance, the relevant knowledge ranged from tokens specific to particular alloys (such as the 80/20 nickel-chromium alloy) through classes of alloys (for example, the family of nickel-chromium-iron alloys) to alloys in general, as exemplified by

k20   Age hardening is a cause of creep resistance.

How well, then, does the reductionist view—that technology generates no new knowledge in its own right and that it merely applies the knowledge born of science—bear up to scrutiny in this particular historical example? We see, in fact, that the invention of the first creep-resistant superalloys entailed both basic science and technological knowledge; and that almost all of the latter in turn was technological (specifically, metallurgical) theory. This theory, in form and nature, is indistinguishable from basic science. But metallurgical theory was generated in the first place primarily because of its practical relevance and usefulness, not for the sake of knowledge itself. It is knowledge about the class of artifacts called alloys that are invented, conceived. and designed rather than discovered or found in nature. To draw on a wonderful distinction made by the metallurgist and historian Cyril Stanley Smith, metallurgical theory is knowledge about material rather than matter.[12] This knowledge emanated from the technology itself, even though it may draw upon physics and chemistry. Furthermore, we have seen that the process of invention of the first superalloys itself entailed the generation of new technological knowledge that *looks* like basic science but is, in fact, metallurgical theory (tokens $k19$ through $k23$). Thus, this particular historical episode yields evidence that the technological process both *uses* knowledge other than basic science and is also instrumental in *generating* new knowledge.

In the case of metallurgical history, Cyril Stanley Smith has made the point that insofar as metallic artifacts are concerned, the basic science of matter has historically been concerned with the structure and properties of atoms and molecules whereas the technology of materials must deal with aggregates, possibly forming hierarchical structures.[13] In the context of structural engineering, Edwin Layton makes a similar observation. "In solid mechanics," he remarks, "engineers deal with stresses in continuous media rather than a microcosm of atoms and forces."[14] Elsewhere, Layton describes this distinction in terms of the transformation of the subject of strength of materials as it originated in physics into a branch of structural engineering:

> As the strength of materials moved from the community of science [that is, physics] to that of technology, it went through an important transformation. Its ties with physics were weakened, and it developed in ways uncharacteristic of the basic sciences. . . . Scientists tended to explain their findings by reference to the most fundamental entities such as atoms, ether and force. . . . The scientists who had done so much to found a science of the strength of materials— notably Young, Coulomb and Poisson—strove to found this study on the same ontological basis as classical mechanics—that is, they sought to explain their results in terms of molecules and the forces between them . . . in the end the attempt was abandoned. Instead engineers were content with a simple macroscopic model—for example, viewing a beam as a bundle of fibers.[15]

Smith's and Layton's insights make the implausibility of the reductionist hypothesis yet more pronounced. They further suggest that the distinction between basic science and technological theory is not only engendered by differences in their very purposes—the former seeking to understand nature, the latter to aid in the construction of artifacts—but also in the fundamental nature of the respective entities of interest. The ontologies of basic science and technological theory are distinct.

Mathematical Knowledge in Technology

Since about the eighteenth century, mathematical principles of one kind or another have also become an integral part of the *formally educated* technologist's knowledge body. As technological knowledge, mathematics is principally an instrument for the building of models and theories, and for analysis and calculation. In some technological domains, what I have referred to above as technological theories are mathematical in form; thus, mathematics and theory are often inextricably interlinked insofar as the roles they play in design and invention.

The general form of this relationship is of the following nature: In order to *solve* a given design problem or to *verify* a solution to such a problem, an abstract, idealized *model* of the situation is constructed. The form of this model is dictated by the available technological theory, or it may necessitate, in the absence of a suitable theory, the development of a new theory. This latter activity constitutes technological research. The model so constructed may be expressible in mathematical terms, for example, by a set of differential equations, a recurrence relation, a directed graph, and so on. The relevant mathematical discipline is then exercised in order to draw conclusions about the model. Such conclusions may help in the solution of the problem at hand, or to verify the correctness of a prior solution.

Technological Knowledge: Operational Principles

When Robert Stephenson embarked on the design of the Britannia Bridge in 1844, and first conceived of the wrought-iron tubular form as a kind of a beam, there already existed an established technological theory of beams. This theory could explain how beams resist bending as a result of vertical loads, and enabled the calculation of internal stresses in beams of different cross sections.[16] However, the available theory was inadequate for computing the ultimate strength, or breaking load, of a beam of some specified material. Furthermore, the experimental data then available applied to cast rather than wrought iron.

Thus, to a large extent, the available technological theory of structures with all its predictive and explanatory powers was of little use to Stephenson. What Stephenson and his colleague William Fairbairn relied upon as the basis of their design and reasoning was a body of knowledge that was neither basic science nor technological theory. Rather, such knowledge was that of how certain kinds of structural forms and structural materials function, behave, perform, and appear under certain conditions.

Following the chemist and philosopher Michael Polanyi, I shall call such knowledge tokens *operational principles*. According to Polanyi:

> Technology teaches only actions to be undertaken for *material* advantages by the use of *implements* according to (more or less) *specified rules*. Such a rule is an operational principle. [italics in the original].[17]

This is, perhaps, too brief for our purpose. Instead, I suggest the following definition; For a given class of artifacts, *an operational principle is any proposition, rule, procedure, or conceptual frame of reference about artifactual properties or characteristics that facilitate action for the creation, manipulation, and modification of artifactual forms and their implementations.*

It might occur to the reader that the output of any design or invention process is itself an operational principle. A design, after all, is a specification of the form or structure of an artifact such that if the artifact is built according to the specification, it is expected to satisfy certain desirable artifactual properties (see Chapter 4). A given design D embodies a relationship between a structure or form, and a set of desired properties as specified by the requirements R. Because designs are themselves operational principles, and since operational principles constitute technological knowledge, the processes of design and invention, cognitively the most significant of technological activities, *necessarily* generates technological knowledge—another blow against reductionism!

But, of course, operational principles also serve as *inputs* to the processes of invention and design. Indeed, they have served as practically the only kind of technological knowledge from the beginning of recorded history until the eighteenth century, when basic science and technological theory began to play more substantive roles. But even in domains that have a theoretical base, operational principles continue to serve as significant sources of knowledge.

Stephenson's and Fairbairn's Knowledge

Consider, once more, the case of the Britannia Bridge and, in particular, the process of ideation that led Stephenson to the concept of the tubular bridge. As noted previously, the suspension bridge—"the most obvious resource of the engineer" as the form of long-span bridges, according to Fairbairn[18]—was initially rejected because "the failure of more than one attempt had proved the impossibility of running railway trains over bridges of that class with safety."[19]

In other words, Stephenson initially rejected the suspension bridge because of the proposition

F_1 Suspension bridges are not sufficiently rigid for the support of rapidly moving railway trains.

This is an operational principle that relates the structure or form of the suspension bridge to a functional property, namely, the ability to withstand a particular kind of dynamic load. It is conceivable that such knowledge may have its origins in technological theory; but given the above quote, it is more likely to have originated from *experience* of the behavior and structural capabilities of suspension bridges.

Later, Stephenson's attention was drawn by J. M. Rendel to Rendel's mode of trussing to prevent oscillation in the platform of suspension bridges. However, Stephenson felt that Rendel's trussing system, while adequate for ordinary roads, would not be strong enough for the purpose of a railway bridge.[20]

In Chapter 7, I gave an outline of how Stephenson may have reached this conclusion by way of a sequence of knowledge-level actions. As part of the process, I suggested the use of the following two knowledge tokens: The proposition (or "fact")

F_6 Road bridges are not generally as strong as railway bridges.

and the rule

R_2 IF e_1 is a kind of structural entity E_1
and
e_2 is a kind of structural entity E_2
and
E_1 is generally less strong than E_2
and
e_1 is designed according to structural style S_1
          THEN   there is no guarantee that e_2 can be designed according to S_1.

Here, F_6 is a comparative statement about two specific artifactual classes ("road bridges" and "railway bridges") with respect to a performance property ("strength"). R_2 is a rule of a very abstract kind—it applies to artifacts in general, not just bridges—that establishes a relationship between structural properties of artifacts, performance, and actions for the creation of artifactual forms.

Both these tokens are instances of operational principles. F_6 *may* have had its origins in theory but is more likely to be a generalization based on empirical evidence or experience. R_2 will almost certainly have been acquired by way of a cognitive act having no foundation either in basic science or theory. And yet, as I suggested in Chapter 7, both these principles may plausibly have participated in the process whereby Stephenson reached his conclusions about the viability of Rendel's trussing system.

The Britannia Bridge also reveals an interesting example of how *technological theory and operational principles interact* during the design process. The reader may recall from my discussion in Chapter 6 that the "later" preliminary experiments, begun in August 1845, entailed among other things the use of a multicellular top to the rectangular form (Figure 6–5). The reason for this modification is explained by Nathan Rosenberg and Walter Vincenti along the following lines.[21]

It will be remembered that the "early" preliminary experiments had revealed failure of the top of the experimental tube by buckling rather than (as had been expected) by crushing. Now, one of the elementary behavioral characteristics of beams is that if a beam is supported at two ends and subject to a vertical load, it bends such that the top surface is in a state of compression and the bottom part is in a state of tension. Under the circumstances, the engineer can conceive the top flange of the tube as an isolated bar or "strut" in a state of compression. Note in passing that here we have yet another operational principle that is a constituent of every structural engineer's knowledge body: to isolate a structural component and treat it as a "free body" with forces acting on it from its environment. Vincenti has described a similar kind of operational principle, called the "control volume," used by engineers dealing with problems involving fluids and heat exchange.[22] Principles such as the free body and the control volume are useful to the technologist since, among other things, they allow abstract *models* of artifacts or components to be conceived, to which technological theory can then be applied.

Now, ever since Galileo's *Dialogue Concerning Two New Sciences* appeared in 1638,[23] it had been known that the load at which a long strut of fixed length and fixed amount of material would buckle can be increased by making the strut in the form of a hollow rather than a solid bar. This effect results from the fact that in a tube or hollow bar, the material is placed farther from the mid line, thereby increasing the rigidity of the strut—to the extent that the strut will not fail by buckling but by crushing of the material. As Rosenberg and Vincenti have noted, it was probably this sort of an idea that prompted Fairbairn to conduct tests on tubes with cellular upper flanges.[24] Citing Edwin Clark, they make the further point that the cellular design of the top flange could be perceived as a set of columns in compression—a structural form with which Stephenson and his colleagues Fairbairn and Hodgkinson were well familiar.[25]

Thus, the knowledge that these engineers brought to bear in arriving at the idea of the multicellular top flange for the tubular beam was, in part, structural theory and the theory of strength of materials (that is, technological theory) and, in part, a model of the top flange as a thin strut in compression (that is, an operational principle). The example also provides insight as to one way by which operational principles and theory can interact in the technological process: operational principles isolate the situation to which theoretical knowledge can be brought to bear.

Newcomen's Knowledge, Revisited

The Britannia Bridge came into existence at a point in history when technological theory and, in some domains, basic science were beginning to make their presences felt. And while it is certainly the case that some technologies are, at any given time, in a more advanced state than others, we may expect, in general, that the further we go back in time, the more pronounced will have been the influence of operational principles as the principle source of knowledge.

It is, thus, of considerable interest, in this light, to reexamine Thomas Newcomen's invention of the atmospheric engine. Recall that Newcomen began this work in 1698 and that the engine became fully operational in 1712. We have also seen that this major event in the annals of technology had a phylogenic history that at the very least can be traced back, on the one hand, to de Caus's toy device of 1615 for raising water by steam pressure and, on the other, to Torricelli's discovery of atmospheric pressure in 1643.

In his biography of Newcomen, L. T. C. Rolt remarked that to the English scientific elite of the eighteenth century, centered as it was around London and the Royal Society, it was unimaginable that a provincial and relatively lowly educated ironmonger—a tradesman—such as Newcomen could have conceived an invention such as the atmospheric engine.[26] Putting aside class consciousness, and the social and intellectual snobbery that may have promulgated such a view, we have seen (in Chapter 8) that despite the inconclusiveness of the evidence, there seems some grounds for believing that Newcomen's engine was the outcome of evolutionary history. It must be reiterated that this in no way detracts from Newcomen's achievement. It was his particular genius to utilize and combine previous ideas and improve upon them to produce the first practical steam engine.

If, in fact, it was the case that Newcomen knew of the relevant work that had preceded his, what sorts of knowledge would he have had? Recalling, once more, the discussion in Chapter 8, we note that Newcomen's engine embodied the following previously invented features:

1. Denys Papin's piston-and-cylinder arrangement of 1691, wherein the pressure of the steam in the cylinder enables the piston to move up, the condensation of the steam produces a vacuum underneath the piston, and the atmospheric pressure on the latter drives it down.
2. Papin's idea of 1675 of covering the piston with water to make it airtight.
3. Thomas Savery's schemes, proposed in 1698, for generating steam in a separate boiler and for raising water from a well by exploiting the atmospheric pressure on the water.

Clearly, each of these ideas constitutes an operational principle in its own right. And though the fact that the atmosphere exerts pressure is itself a piece of basic science discovered and explored by such natural philosophers as Evangeliste Torricelli, Otto von Guericke and Christian Huygens, and though there is indeed a link between such science and the later invention, the science itself need not be considered to be a *necessary* component of Newcomen's knowledge in order to explain his inventiveness. Rather, if the evolutionary account is correct, Newcomen's knowledge was the knowledge of the operational principles enunciated above.

We can appreciate this more if we consider, for example, Papin's invention of the piston and cylinder arrangement—and keep in mind that all artifactual designs, by their very nature, are operational principles, not technological theories. Papin's scheme exploited two *natural* phenomena: the force exerted by the atmosphere, as discovered by Torricelli in 1643; and the expansive power of steam. Concerning the latter, Papin was also aware that water when turned to steam acquired "an elastic force," which disappeared when the steam was condensed back into water.[27]

This phenomenon was, apparently, well known by Papin's time, according to the historian of the steam engine, H. W. Dickinson, who cited the laboratory experiments carried out by the natural philosopher Giambattista della Porta (1538–1615)[28]—an Italian "of limitless curiosity"[29]—and described in his book *Spiritali* in 1606. Thus, Papin's own invention in 1691 was founded on basic sciences or, as it would probably have then been called, "experimental philosophy." Papin thus concluded that this property of water-steam could be exploited, with the expenditure of very little heat, to produce a "perfect" vacuum.[30]

Papin's piston-and-cylinder arrangement, thus, assimilated his physical knowledge, of the expansive power of steam and of the phenomenon of atmospheric pressure, into an operational principle. Newcomen had no need to be aware of Torricelli's discovery or della Porta's experiments. These had been taken and remolded into artifactual principles.

Operational Principles in the Invention of Computing Devices

The term "high technology" has come to refer to the most sophisticated forms of devices and manufacturing processes, which are, in some way or another, associated with the computer on the one hand, and the application of basic science on the other. The computer itself is the flagship of high technology. And indeed, if we pause to reflect on the origins of the early mechanical, electromechanical, and electronic computers, we realize that perhaps no other class of artifacts in the history of technology has been sired by a more educated and scientifically trained group of people. Practically everyone of the principal figures in the development of the computer, including Alan Turing, John von Neumann, Howard Aiken, John Mauchly, Grace Hopper, Arthur Burks, John Atanasoff, Jay Forrester, Maurice Wilkes, Frederick Williams, and Tom Kilburn, taught and did their research at universities, mostly holding Ph.D. degrees in physics, mathematics, or electrical engineering. Indeed, we should not forget that the most celebrated pioneer of the digital computer from the Victorian era, Charles Babbage, was the Lucasian Professor of Mathematics at Cambridge University—a chair once held by Sir Isaac Newton and occupied in our own time by Stephen Hawking! One would expect, then, that their knowledge of basic science, mathematics, and technological theory collectively played a vital role in their respective contributions to the development of the computer. And yet even in this domain, we find that operational principles constitute a significant, even major, component of the knowledge individuals have summoned for their respective causes.

Consider, for example, John Atanasoff's conception of the add-subtract mechanism (ASM) for the Atanasoff-Berry Computer (ABC). Atanasoff, a Ph.D. in theoretical physics from the University of Wisconsin in 1930, and holder of degrees in electrical engineering (B.S., 1925) and mathematics (M.S., 1926) was, at the time he began to think about his machine in 1937, deeply versed in electronics as well as the state of the art of both analog and digital computers. One of the key decisions he made regarding the ASM was that it would perform its computations by "direct logical action and not by enumeration as used in analog calculating devices."[31] He had also decided to use a base-two (or binary) system for his machine. In this context, Atanasoff was drawing upon his *mathematical* knowledge of number systems other than that of the base ten—knowledge that led him to explore a range of alternative bases for his machine and which enabled him to analyze in some depth the advantages of using base two.[32]

However, neither basic science nor technological theory was helpful as a source of ideas as to how precisely to implement "by direct logical action" the ASM unit. Had this situation arisen a few years later, a digital designer would have drawn upon Boolean algebra as the theoretical or mathematical basis. Atanasoff, apparently, did not recognize the applicability of Boolean algebra to his problem—it appears that he was unaware of Claude Shannon's landmark master's thesis at the Massachusetts Institute of Technology on the application of Boolean algebra to the analysis of switching circuits, work that was published in 1938.[33] It is interesting to note, though, that Samuel H. Caldwell, who was on Shannon's M.S. thesis committee, had corresponded with Atanasoff during 1939 and 1940 and had actually visited Iowa in the fall of 1939 to examine and review Atanasoff's project on behalf of the National Defense Research Corporation.

Instead, Atanasoff drew upon the idea of a computing circuit that would operate on a digit-by-digit basis, from the low-order (units) position toward the higher order end, wherein at each digit position, the circuit would take the two input digits and a carry from the previous or next lower order position. This was how the abacus, that most ancient of mechanical calculating devices, worked in performing addition and subtraction—and, indeed, according to his biographer Clark Mollenhoff, Atanasoff initially visualized the ASM as operating on a base-two system, on the principle of the abacus.[34] In his definitive 1940 memorandum on the design of the ABC, Atanasoff does not mention the abacus although the two primary memory inputs to the ASM were named "keyboard abacus" and "counter abacus" respectively.[35] Rather, he there described the arithmetic circuit operating according to "principles that are analogous to the function of the human brain in mental calculation."

It seems reasonably safe to say that the process of ideation that led to the arithmetic unit in the ABC relied on knowledge that was, despite Atanasoff's considerable background in science and engineering theory, the shared principles of calculation performed on the one hand by the abacus and, on the other, by humans. In this case, the principle specified a *procedure* or *algorithm* for effecting the arithmetical *functions* demanded by the artifact.

Maurice Wilkes, like Atanasoff ten years his senior, was trained as a mathematician and physicist. Wilkes obtained his bachelor's degree in mathematics—the celebrated "mathematical tripos"—from Cambridge in 1934 and his Ph.D. in radio physics in 1938. Yet when we examine the circumstances attending his invention of microprogramming, we note above all else that the two dominant knowledge tokens that contributed to his process of ideation were the principle of the stored-program computer and the circuit form called the diode matrix. The former is an operational principle that stipulates how the structure and behavior of a device yields a flexible computational capability;

the latter relates a highly organized circuit form to the desired property of structural regularity. And while the diode matrix as a circuit is itself grounded in theory, namely, the physical theory of diode behavior and the theory of digital circuits, such issues are abstracted away, leaving only the operational principle of the diode matrix. It was this principle, not the theory underlying it, that Wilkes drew upon in conceiving the idea of microprogramming.

Leaving aside these two "grand" principles, if my account of Wilkes's process of ideation[36] is plausible, then many other instances of operational principles are revealed. Referring to my discussion of this invention in Chapter 7, the fact

F_5 In the MIT Whirlwind, each arithmetic operation (barring multiplication) involves exactly eight pulses issued from a control unit implemented in the form of a diode matrix.

relates the structure to the behavior of a *particular* artifact. The assertion

F_7 Whirlwind's control unit is simple, repetitive, and regular.

stipulates structural and aesthetic properties, also of a particular artifact. The rule

R_2 **IF**   an order interpreter is (partially) implemented using a diode matrix
       THEN   (that part of) the order interpreter is simple, repetitive, and regular.

relates a set of subjectively perceived but desirable properties ("simple, repetitive, and regular") to a particular kind of artifactual form ("diode matrix").

Theory versus Operational Principles in Software Technology

Software, as a class of complex artifacts that is deeply associated with but distinct from, the physical computer or "hardware," may be said to have come into explicit existence with the very first "operating systems"—computer programs that are responsible for the management of hardware resources when a computer is in operation. Such operating systems, albeit of rudimentary form, were first conceived in the late 1950s and early 1960s. Since then, not only have operating systems grown into massively complex, multifunctional entities, but innumerable other kinds of software artifacts of comparable structural and functional complexity have emerged.

Historically speaking, neither basic science nor technological theory helped in the development of the earliest software systems, simply because there was no science or theory that could be applied. Software designers of the time— and the word *design* was then scarcely in use in this context; one simply "wrote" software—invented operational principles as and when required. Researchers in academic settings consciously formulated new rules as canonical principles for good design. In this sense, the technology of software was as craftlike as the art of the wheelwright described so beautifully by George Sturt[37]—except that the early software designer had not the benefit that Sturt's wheelwright had of centuries of experiential knowledge.

The materials out of which software-as-artifact is created are, of course, the symbol structures we call computer programs. And, between 1967 and 1969, two computer scientists, Robert Floyd in the United States and Anthony Hoare in Britain, wrote the first papers that drew upon the formal mathematical nature of such symbol structures and established a mathematical basis for the design of correct programs. Thus began the development of a *technological theory* of software that has grown into a formidable body of knowledge concerning the logical and mathematical relationship between the structure and the function of computer programs.

In practical terms, the theory of programming relies on a collection of axioms that define, mathematically, the behavior of the "atomic" concepts from which programs are built (such as the most elementary statements in a programming language); and rules of inferences that allow one to deduce, logically, the behavior of more complex software entities such as a subroutine, a program module, or an entire program from the behavior of its constituent elements. Thus, given a specification of requirements that the program is intended to satisfy, these requirements also being stated as mathematical expressions or formulas, the theory can be used to prove that the behavior of the designed software system meets the requirements.

The mathematical theory of programs, then, plays a role in the design of software that is analogous to the role played by the theories of strength of materials and structures in the design of structures by civil engineers: each is a tool for analyzing the relationships between structure, function, and behavior of the relevant class of artifacts.

But here lies the conundrum: Despite the rapid growth of the theory of programming as a body of knowledge in the quarter century since its birth—despite the considerable level of communication of the theory, by some of the most respected scientists in the field of computing, through the medium of articles and books—the theory and its development have been virtually ignored by software practitioners. The practice of software engineering has largely been pursued with virtually no reliance on the theory.

One of the most widely read and referenced tracts on software engineering is *The Mythical Man Month* by Frederick Brooks, a former designer and technical manager of one of the most significant computer system design projects of the 1960s and, later, a professor of computer science. Admittedly, the book was published in 1975. Yet it is worth noting that though Brooks discusses one of the key concepts concerning the software design process, a concept called "structured programming," which had emerged only the year before, he makes no reference whatsoever to the mathematical theory of programming.[38]

As another more recent example, in *Software Perspectives*, Peter Freeman, an academic computer scientist and software consultant, presented a cultured view of the software design and development process with not a mention of the mathematical theory of programming.[39]

Still more recently, the computer scientist Mary Shaw, who had been closely associated with the influential Software Engineering Institute in Pittsburgh, writing on the "Prospects for an Engineering Discipline of Software" and, in particular, on the possibility of a scientific foundation for software engineering, makes only the briefest of mentions of the mathematics of programming: "The 1970s saw substantial progress in supporting theories, including performance analysis and correctness."[40]

Perhaps the most telling evidence of the extent to which software development seems to have been conducted independent of its theoretical basis is Donald Knuth's description of the process by which his highly innovative computerized typesetting systems TEX and METAFONT were developed. As one of the most influential persons in the field of programming from the mid-1960s to the mid-1980s, Knuth was intimately acquainted with and actually contributed to the mathematical theory of programs.[41] Regarding TEX, Knuth was to write: "TEX was the first fairly large program I had written since 1970; so it was my first nontrivial "structured program" in the sense that I wrote it while consciously applying the methodology I had learned in the early 70s from [E. W.] Dijkstra, [C. A. R.] Hoare and [O.-J.] Dahl, and others. I found that structured programming greatly increased my confidence in the correctness of the code, while the code still existed only on paper."[42]

Thus, Knuth apparently applied the principles of structured programming in the course of developing TEX. Yet there is no mention whatsoever, in his otherwise ruthlessly candid dissection of his progress through the project, that he was influenced by, or that he summoned to use, the mathematical theory itself.

My hypothesis, then, is as follows: *The predominant kind of knowledge that the software designer brings to bear in the creation of software artifacts is not theory. Rather, it is constituted of operational principles.* Many of these operational principles—the most significant of them, certainly—are highly abstract and conceptual in nature, but they are by no means exact in the sense of the basic sciences, technological theory, or mathematics.

The extent to which operational principles constitute the dominant kind of knowledge in software engineering is best exemplified by considering a single short period in the history of computing, 1970–75, when the new Kuhnian paradigm of software design called the *structured programming paradigm* came into being. Echoing Thomas Kuhn, computer scientists and software engineers often refer to the advent of this paradigm and its aftermath as the structured programming revolution. Much as Kuhn had described the nature of scientific paradigms in general, the structured programming paradigm had several interrelated components to it. And as in successful revolutions in science, the structured programming revolution—a technological rather than scientific revolution—has yielded a working paradigm which subsequent practitioners of software engineering from the 1980s on have taken for granted. Its elements are now part of the knowledge body of most software designers. The striking aspect of the paradigm is that although there are indeed some theoretical components to it, the operational principles components are the ones that have been most widely assimilated and used.

The Ontology of Operational Principles

The examples given in the foregoing sections—and many others could as well have been cited—illustrate the ubiquity of operational principles in technological creativity. The mere fact that I have devoted so many pages to its discussion is indicative of the importance that this writer, at any rate, attaches to operational principles as technological knowledge. In fact, I will suggest that if we wish to seek the common element, if any, between technologists through historical time—to seek, that is, what it is that links the metalworkers of preclassical times, the builders of medieval cathedrals, the inventor-enterpreneur of the Industrial Revolution, the bridge builder of the Victorian age, and the present-day exponent of "smart" automata—it is to the nature of their knowledge that we must look; and while the knowledge that technologists have brought to bear on their respective domains will, in part, be obviously specific to that domain, and may include basic science, mathematics, and/or technological theory, it is to the knowledge of operational principles we must look for the shared element. This leads me to frame

> *The operational principles hypothesis*: The essence of what makes a person a technologist is that his or her cognitive act of creativity is driven by his or her knowledge of operational principles.

Homo sapiens became *Homo faber* not by virtue of his familiarity with nature's laws—though this helped—nor his adroitness in the manipulation of abstract symbol structures—which also helped—but by way of his intimacy with the artifactual form itself, its variety of features and properties, and by knowing or having the capacity to hypothesize how forms with such desired properties or characteristics can be achieved. It is such knowledge which, following Michael Polanyi's suggestion, we are calling operational principles.

Once we recognize this epistemic character of technology, some heretofore puzzling aspects of the nature of the technological process are clarified. It explains, for example, how it is possible for remarkable acts of inventiveness to have been achieved through history by artisans and craftsmen who, according to conventional wisdom, are thought of as being "unlettered," "lowly stationed," or "uneducated." Operational principles fall neither in any of the categories of the liberal arts of ancient Greece nor in what used to be called natural philosophy. Hence, according to the orthodoxy, they do not constitute knowledge. They are dismissed, to the extent they are pondered upon at all, as "ad hoc rules" or "rules of thumb." It is to the eternal credit of the pioneers of artificial intelligence, Herbert Simon and Allan Newell—and among their most formidable intellectual achievements—to recognize that operational principles, in the guise of their term, *heuristics*, constitute the most ubiquitous kind of knowledge used in human problem solving.[43] And what is technology but the solving of problems, by humans, pertaining to the creation of artifacts?

Light is also shed on the questions I posed in the Prolegomenon: How did the scientifically illiterate smiths in the shanty huts across from the campus where I had studied as an undergraduate forge red-hot steel so perfectly, knowing nothing about heat treatment or phase transformations in alloys? What did *they* know of metallurgy that we did not, schooled though we were in the most advanced theories of solid-state behavior?

What *I* knew (or thought I knew) of metallurgy was, mostly, the basic science of solid-state physics, and the technological theory of phase transformations in metals and alloys. What *they* knew were the unwritten qualitative rules correlating the color of heated metal to its malleability and forgeability. What *they* knew were operational principles. Indeed, as far as their craft was concerned this was, most likely, their only kind of knowledge.

The notion that operational principles constitute the primary kind of technological knowledge was, I have mentioned, first articulated explicitly by Michael Polanyi in his wide-ranging tract *Personal Knowledge*.[44] More recently, also influenced at least in part by Polanyi, the social scientist Donald Schön, in his study of how professionals—doctors, lawyers, architects, engineers—think in the practice of their vocations, made the distinction between "technical rationality" and "reflection-in-action."[45]

Technical rationality is the notion that professional practice "consists in instrumental problem solving made rigorous by the application of scientific theory and technique."[46] Historically, technical rationality has its roots in the doctrine of positivism, expressed first by the nineteenth-century French thinker Auguste Comte. This doctrine suggested, first, that empirical science constituted the only source of genuine or "positive" knowledge in the world; second, it advocated that "men's minds should be cleansed of mysticism, superstition and other forms of pseudoknowledge;"[47] and third, that scientific knowledge should be extended and applied to the domain of human affairs.

Positivism was transformed in the first half of this century, by way of the famed Vienna Circle, into the standard philosophy of science. Knowledge or meaningful propositions, according to this philosophy of "logical positivism," consists of either the analytical, tautological propositions of logic and mathematics, or the empirical, testable propositions that describe nature. But, Schön pointed out, professional knowledge—the practical knowledge brought to bear by the professional—seemed to fit into neither of these categories. Positivism, according to Schön, solved this problem by regarding practical knowledge as a rational issue that relates means to ends. The question of how one must act in order to achieve some practical end becomes, for the practitioner, a matter of selecting the best means to adopt for this purpose according to what science advocates.[48]

In the context of technology, technical rationality becomes, of course, the doctrine of technology-as-applied-science. A serious problem with technical rationality as it applies to both human affairs and the world of artifacts is that it reduces professional practice to problem solving—that is, the determination of means to ends—on the assumption that the ends are fixed or known. But when the ends themselves are unclear, imprecise, or incomplete—when, to use Herbert Simon's term, problems are *ill-structured*[49]—then the doctrine of technical rationality fails, for science is unable to resolve questions about goals or ends. This is not the only limitation to technical rationality, as Schön points out, but it is sufficiently important a problem itself to undermine its authority. For, in the context of the artifactual world, in the context of design, a substantial component of the design process itself is concerned with the transformation of ill-structured problems or goals into ones that are precise or complete or "well-structured."[50] And, as I have shown elsewhere, such a transformation, in the majority of cases, will appeal to rules, hypotheses, or conjectures having nothing to do with basic science or technological theory.[51] For example, in designing a user interface for a personal-computer-based workstation, if an initial goal is

The user interface must be easy to use.

since there are no absolute objective criteria for "ease of use," the designer may appeal to such a rule as

IF   the means of communication is natural language
THEN   the user interface is easy to use.

This is clearly an instance of an operational principle.

Schön's response to the failure of technical rationality in the domains of professional practice is to propose that identifying, defining, and solving problems in these domains entail what he calls *reflection-in-action*.[52] While Schön's theory of reflection-in-action is not easy to state in precise terms, its main characteristics can be identified as follows:

1. The knowledge that the practitioner brings to bear is often *tacit* knowledge—knowledge that is hard to describe or can only be described inadequately.[53] Such knowledge is only expressed by virtue of actions performed. Schön refers to such tacit knowledge as *knowledge-in-action*. The classic example is the knowledge that a person utilizes when riding a bicycle. The knowledge lies in the very act of riding. There are no expressible rules.

2. Professional practitioners also *think* about what they are doing in the very process of doing it. Schön refers to this as *reflection-in-action*. It entails, for example, making *conscious* adjustments to one's actions "on the fly," based on feedback from the situation at hand and from the outcome of a prior action. As an example, Schön describes the way jazz musicians who are improvising together make on-the-spot adjustments to their music based on the sounds that they hear.[54]

3. The professional practitioner practices. And in the conduct of her practice, she resorts to a repertoire of "expectations, images and techniques." She learns what to look for, and in any given situation she looks for such

elements. Indeed, the routine element of professional practice entails the mapping of the problem that the practitioner is posed onto one or more of her repertoire of "cases." Schön refers to this kind of knowing and its application as *reflecting-in-practice*.

The main point of Schön's argument is that none of these modes of knowing conform to the model of technical rationality. And while I disagree about the essential ineffability of what Schön refers to as knowledge-in-action—we *think* we cannot express the knowledge entailed by acts of walking, for example, or gripping things, and yet such knowledge is being explicitly embedded in the design of robots—it is certainly the case that much of the kinds of knowledge described in the concepts knowledge-in-action, reflection-in-action, and reflecting-in-practice, are not part of basic science or technological theory. They are precisely the kinds of knowledge I have described here as operational principles.

How Do Operational Principles Originate?

How are operational-principles-as-knowledge produced? We have a reasonably clear understanding of the manner in which scientific or mathematical knowledge, or technological theory is generated—the basic mechanism involved is the explicit, mostly conscious, human activity we call *research*. But what of operational principles?

Design and Invention as Sources of Operational Principles

The most explicit means for the production of an operational principle are the processes of design and invention. Recall that the outcome of every act of design is the delivery of an artifactual form which, when built, will satisfy—or is intended to satisfy—a given set of attributes called the "requirements." A design thus embodies and encapsulates an operational principle. Insofar as the goal of design is to produce an artifactual form that has not existed previously, every design embodies or encapsulates (at least nominally) a novel operational principle. In the case of invention where the artifactual form is historically novel or historically original, the operational principle is also historically novel or original. Here we have an explicit twofold connection between the acts of discovery and invention: On the one hand, both generate new knowledge; on the other, one generates theoretical and scientific knowledge while the other produces operational principles.

Experiments as Sources of Operational Principles

In natural science, experiments—that is, observation and measurement of phenomena in a controlled setting—serve to test hypotheses. In the event an experiment corroborates a hypothesis, it serves to consolidate a piece of scientific knowledge. If it falsifies the hypothesis it is also instrumental in producing knowledge, albeit of a negative sort. In technology, experiments also obviously serve similar purposes: as the basis for the testing of hypotheses and the generation and consolidation of technological theory; and for the generation of new hypotheses. Thus, for example, the Mond metallurgists' use of X-ray diffraction studies of the nickel-chromium alloys led them to conclude that such alloys contained an iron-based phase known as "sigma"—a piece of technological theory that, as it turned out, was actually false.

But experiments of the kind that scientists are familiar with also yield technological knowledge in the form of operational principles—especially in the absence of theory. In the case of the Britannia Bridge, a great deal of knowledge that this project yielded was of the operational kind; and much of this, in turn, was generated directly from experiments. Specifically, the early experiments with the various tubular forms produced the following knowledge token:

> The rectangular tube section is superior to elliptical and circular sections in its resistance to distortion.

This hypothesis was not deduced from any theoretical premise; at the same time, it served as the basis for Stephenson and Fairbairn's decision to adopt a rectangular section for the tubular bridge. It was an operational principle derived from experiment. Likewise, the later experiments produced knowledge of how the buckling of the top flange of the tube could be eliminated by adopting a corrugated cellular form for the top.

The scale-model experiments carried out later were of two types. One set was performed by Eaton Hodgkinson, a theorist, to determine the ultimate strength of tubes of different dimensions,[55] and to generate further basic knowledge of plates and tubes in compression and bending.[56] Such knowledge, we could say, belongs to technological theory. The other set was conducted by William Fairbairn, a practitioner, to determine the ratio of the cross-sectional areas of the top and bottom flanges so that they would fail simultaneously. These experiments were conducted in the context of a *specific* artifact and yielded such knowledge as that the top and bottom flanges fail simultaneously when the ratio of their areas is about 1:1. It also produced, partly as a consequence of the above result, the final hypothesis:

> A rectangular sectioned tubular wrought iron bridge with a cellular top . . . and a cellular bottom . . . will satisfy the strength, stability and navigational requirements.

Experience as Source of Operational Principles

It is a commonplace that a large part of our conceptual understanding of and beliefs about the external world is obtained through direct experience. It is no less so even in the more structured world of the technologist. Experiential technological knowledge is simply the repertoire of empirical facts, hypotheses and rules that are assimilated into an agent's knowledge body as a result of direct observation of phenomena pertaining to artifacts or other entities. And while such experiential knowledge may lead to subsequent theoretical knowledge—as a result of deeper, more systematic analysis of phenomena—in essence, such knowledge is composed of operational principles. Experience is, thus, a major source of such types of knowledge.

Much of the operational knowledge the craftsman of old possessed was no doubt experientially gained. George Sturt in *The Wheelwright's Shop* wrote memorably about the phenomenon of "dish" in wagon wheels: The dish of a wheel lends it concavity outwards, a bit like a plate or a saucer (Figure 9-1). Sturt pointed out that in the course of learning the wheelwright's trade he found that wagon wheels should have dish, for otherwise, through use, a wheel would "turn inside out, like an umbrella in a gale."[57] But he could not fathom why this was so. Why should a load "if really too heavy, sometimes crumble a wheel up" in that way rather than others?[58] His understanding of the "why" came about one day by his observation of the pattern of the road surface produced by horse traction:

> It will be remembered that, in the old days of horse-traction, roads were worn down, crosswise, in a wave-work of shallow hollows from one side to the other, about a yard or so apart. . . .
> . . . One day I saw how this road surface was produced. A cart before me was gently swaying from side to side with every step of the horse. And as it

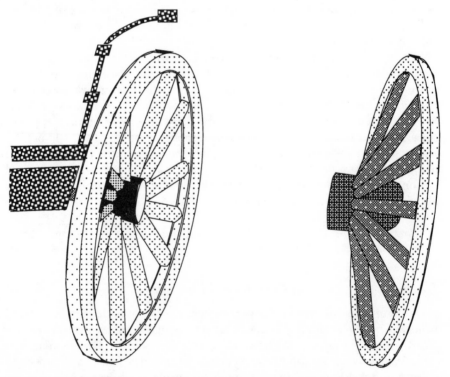

Fig. 9-1. *The "dish" or outward concavity in a cartwheel* (Adapted by permission of the publisher from G. Sturt, *The Wheelwright's Shop*, p. 93, © 1993 Cambridge University Press). This characteristic enables the wheel to offer resistance to the continuous lateral thrust of the cart against the center of the wheel.

swayed, the wheels—first the off, then the near, in fine alternate rhythm—loosed a grain or two of the road-metal, ground the channel across the road, a grain or two deeper, now at this step, then at that. . . . As long as . . . horses trotted one step after another from side to side slightly swaying, so long the carts and waggons . . . behind the horses would gently grate into the road surface and wear it into those wary channels."[59]

It was then that he realized the reason why wheels need dish: "The loaded body of cart or wagon, swinging to the horses' stride, becomes a sort of battering ram onto the wheels, this side and then that. . . . And so it goes on with every horse, all day long. The wheels have to stand not only the downward weight of the load; a perpetual thrust against them at the centre is no less inevitable."[60]

The dish of a wheel, Sturt later remarked, is no whim that could have been avoided. It was in response to "a law as inevitable as gravitation."[61] And this "law" was that, in addition to the vertical load (or "downward weight of the load"), there was a lateral thrust ("sideways push") against the center of the wheel. Such is the nature of experiential knowledge.

Experience, of course, includes the experience and the recording of *failure*. The outcome of such experience is very often new knowledge. One learns from one's mistakes. For otherwise, to paraphrase the philosopher George Santayana, those who do not remember past errors are condemned to repeat them.

Perhaps the most interesting form of knowledge that the experience of failure yields is the identification of different types of *errors* and of relationships between errors and the failures that they may cause. Again, such knowledge often, indeed most often, cannot be mapped onto basic science or technological theory. Such knowledge is not always exact. Rather it takes the form of general categories or rules. Henry Petroski, for example, on the basis of his examination of a series of historical cases of errors and failures in structural design, identified four different classes of errors.[62] For each effort type he also gave what he called a *paradigm*—a word that he uses in the pre-Kuhnian, dictionary sense to mean an archetypal example.

One of Petroski's error types is what he called "flawed conceptual design" but which more accurately should perhaps be called the "error of inadequate criticism of a conceptual design." His example of this error type is a scheme devised in classical times by a Roman contractor named Paconius to move a massive stone pedestal from a quarry to its destination—a temple of Apollo. Paconius's scheme was an ingenious modification to an earlier scheme for transferring columns and other kinds of stone blocks from quarries to building sites—the modification being needed because of some design constraints on the transportation of pedestals that were not present in the case of other kinds of stone blocks. However, so confident was Paconius about his design of the transportation process that he did not criticize it sufficiently nor carry out tests on a small-scale model of the scheme.[63]

Petroski's other error types include the error of "oversimplified assumption" for which his exemplar is Galileo's celebrated but flawed analysis, in his *Dialogues Concerning Two New Sciences*, of the strength of cantilever beams,[64] and the error of "misunderstood scaling effects," examples of which are also taken from Vitruvius and Galileo.[65] Finally, Petroski has discussed the Britannia Bridge itself, despite its structural success and originality at the time, as an example of the "rubric of tunnel vision"[66]—in the case of the Britannia Bridge this being the error of focusing too narrowly on the structural problems and ignoring other issues such as economic and environmental problems.

The extent to which the experience of failures and errors yields knowledge has also been revealed by Donald Knuth in his description of the errors generated in the course of his design and implementation of the TEX computerized typesetting system.[67] Knuth identified thirteen different categories of errors. These included, for example, errors of omission, wherein an intended action or operation was not actually performed, and the misuse of a programming-language feature through misunderstanding of the language or system hardware capability.

It is important to note that the errors that Petroski identified from his historical studies and those that Knuth discovered more directly in the course of design cannot be categorized as belonging to basic science. Nor are they formal enough to be mathematically expressed. Nor can they be considered to be elements of the relevant technological theory. In fact they belong, properly speaking, to the psychological domain; in one way or another they reflect the limitations of agents' cognitive capabilities. As such, viewed as technological knowledge, they can only be considered as rules or categories that can guide the technologist in his or her actions.[68]

Operational Principles Abstracted from Science and Technological Theory

We have seen that operational principles can originate in the absence of scientific or theoretical understanding of a technological phenomenon. This is how technology has primarily developed in the course of history till relatively recent times. Indeed, to a large extent, the role of basic science and technological theory has been to provide explanations of operational principles that may have been identified for years, even centuries or millenia before. A truly fascinating example of this role of science and theory in technology is Brian Cotterell and Johan Kamminga's analysis of the mechanical underpinnings of preindustrial technology, in which they explain the operational principles of many ancient artifacts, including machines, structures, tools, and musical instruments, in terms of Newtonian mechanics.[69]

There is also, however, the obverse side of the coin. Technologists in the course of their practice do not always resort to basic science or technological theory as the source of their immediate knowledge, even when such knowledge exists. Very often, such "deep" knowledge is abstracted into the form of convenient operational principles. Such principles are then derived from, and are abstractions of, science and theory.

Operational principles can and do work in tandem with science and theory. A given operational principle based on empirical studies, observation, or experience prompts research into its underlying causes. The resulting scientific and theoretical knowledge in turn generates, through compilation and abstraction, new operational principles.

Several examples of this phenomena can be offered, especially from those technologies having a long pedigree, but two will suffice for our purposes.

In the domain of civil engineering, the ancient Romans used *arches* quite extensively in building their bridges and aqueducts. Stephen Timoshenko has pointed out that, so far as is known, there was no theory to guide the arch builders of that time. The Romans were dependent only on "empirical rules."[70] Thereafter, interest in arch building lay dormant through the Middle Ages but was renewed during the Renaissance. Even during this time, arch designs were based on empirical rules. The arch designer's knowledge both in Roman and Renaissance times was, thus, composed of operational principles.

A theory of arches was born only toward the end of the seventeenth century when a Frenchman, Philippe Lahire (1640–1718), began to apply the laws of statics and geometry to the analysis of arches.[71] His book *Traite de Mechanique*, published in 1695, and a memoir published in 1712 in *Histoire de L'Academie des Sciences* thus established the beginnings of a technological theory of arches. If an arch is conceived as a successive set of adjacent, identical wedge-shaped segments (Figure 9-2), then Lahire's theory pertained to the weights that these segments must have in order to guarantee that the structure is stable. His analysis revealed that stability is ensured if the weights of the wedges are in a certain ratio. Using the same analytical technique, Lahire also determined the requisite dimension of the pillars that would support a semicircular arch.[72]

We thus see a transition from a traditional knowledge body of operational principles (the "empirical rules") to theory. But, then, this theory was subsequently applied by others in France to prepare a *table* for the purpose of calculating the thickness of arches.[73] Such a table becomes an abstraction of the theory; it constitutes a new operational principle.

As a second example, one of the established methods for manufacturing metal products is a technique called *powder metallurgy*. The practice of this technique has been traced back to Egypt in 3000 B.C. The famous Delhi Pillar (Figure 9-3), weighing approximately six tons was also, apparently, made by this technique in India in A.D. 300.[74] In its modern form, the technique was first applied in the early part of the nineteenth century to produce platinum ingots. The process in essence consists of three steps: (a) The production and treatment of metal powder; (b) compaction of the powder mixture at a suitable temperature and pressure; and (c) sintering the compacted powder—that is, heating the compact to very high temperature in special furnaces under strict atmospheric conditions so that the pores between particles gradually contract and eventually disappear— in order to develop in them suitable mechanical and other characteristics.[75]

Prior to about 1920, sintering was conducted on the basis of trial and error or rules of thumb—that is, on the basis of certain kinds of operational principles.[76] Subsequent systematic studies established that the physical nature of the sintering process involve certain mechanisms known in physical metallurgy, namely, diffusion, crystallization, and grain growth.[77] The elucidation of these

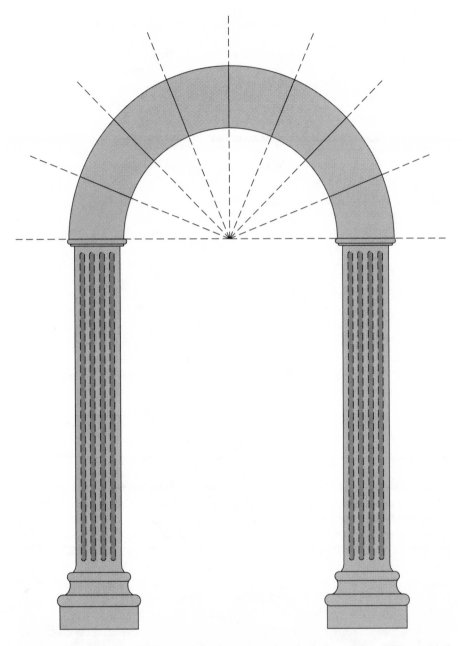

Fig. 9-2. *An arch viewed as a set of adjacent wedge-shaped segments.* This enabled Philippe Lahire to apply the principles of mechanics and geometry and thereby analyze the stable behavior of such arches. Lahire's theory was then codified into a table for the purpose of calculating the thickness of arches.

mechanisms thus constituted the emergence of a technological theory of sintering. The theory, in turn, yielded new guidelines—operational principles—for more effective sintering.

Operational Principles as Pictures

In their monograph on the Britannia Bridge, Rosenberg and Vincenti explain the notion of buckling in thin long struts or columns in terms of what happens when a drinking straw is pushed at its two ends.[78]

Here, they are relying upon two kinds of things to explain buckling. First, the reader's experiential knowledge of what happens when one pushes on the

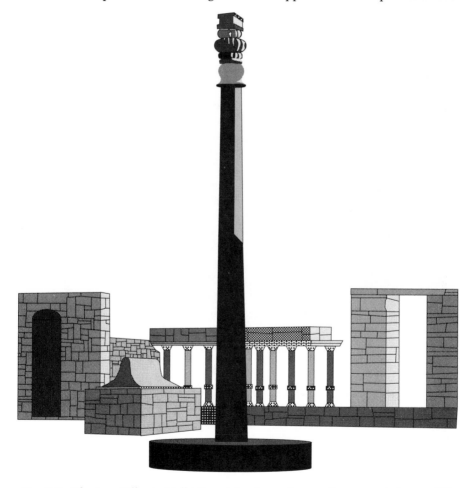

Fig. 9-3. *The Iron Pillar in Delhi.* It weighs about six tons. It was made in A.D. 300, apparently by means of the technique of powder metallurgy—an ancient technology known at least as far back as 3000 B.C. in Egypt. A scientific understanding of powder metallurgy, however, has emerged only in the past century.

two ends of a drinking straw—that is, the operational principle of the behavior of a straw on which forces are imposed at the ends; and second, *the reader's capability to imagine* the way that the straw buckles. For when I read this passage, I do not at the conscious level conjure up rules or networks of ideas. Rather, I see a mental picture of the straw stood vertically on a table and pressed down from the top. I see, in my mind's eye, the straw folding up, *buckling*, as a result of the pressure. I seem to resort to what cognitive psychologists would call a *mental model* for the buckling behavior of the straw.[79] Of course, when such a mental picture is to be described or communicated one can resort to textual means. But one can, as well, actually *draw* the image.

If, as has been suggested here, operational principles form the dominant types of knowledge that the technologist resorts to when designing or inventing, then we can begin to appreciate the point advanced by many engineers and other observers that technological thinking is by and large *nonverbal* or *visual* in nature. This view has perhaps been articulated most convincingly and thoroughly by the historian of technology and former engineer Eugene Ferguson in his recent work, *Engineering and the Mind's Eye*.[80]

We can understand why Ferguson and so many others before him have thought this to be the case.[81] For, as already noted, the output of every act of design or invention is an operational principle describing the form and, by inference, behavior and operation of some desired artifact. An artifactual form—an arrangement of components—and its behavior lend themselves most naturally to the construction of pictorial mental models—that is, mental representations of what the artifact will look like and how it will behave and operate in the real world. Such mental models have the added advantage that the agent can actually conduct "thought experiments" and simulate the internal representation. Ferguson gives the example of the nineteenth-century inventor of the steam pile driver, James Nasmyth, who by his own account would construct in his mind the structure of a machine, "set it to work," and "observe" its performance as if it existed in the material world.[82]

Yet another signpost of the significance of visual thinking in technology is the fact that the primary mode of describing the product of design is the *engineering drawing*. Indeed, we take this so much for granted that there is a common tendency to refer to the engineering drawing of an artifact as "the design."

It is useful to keep in mind that technology has not always relied on such external representations. For example, as Christopher Jones has noted, the traditional craftsmen did not externalize their designs or ideas through drawings or other representational means. Craftsmen held their knowledge and their ideas in their minds and such knowledge was transmitted orally, or was acquired by their pupil-apprentices through observation. External representation in the form of drawings only came about when a given type of artifact became too complex for the cognitive capacity of the individual craftsman; it also arose when the demands of productivity necessitated the cooperative work of many craftsmen who then needed to communicate with one another through a shared medium.[83] Finally, whenever and wherever there was a separation of *conceiving* the artifactual form from its *making*—whenever, that is, designing and implementation became separate tasks involving separate groups of men—the drawing was the means for the design to be communicated to, and as a source of instruction for, the artifact maker.

In any case, for such artifacts that necessitated external representation in some symbolic form, the preferred form for such representation has been the drawing. Furthermore, as Ferguson has pointed out, such drawings are not merely drawings of the final design. Designers also use informal drawings or *sketches* during the course of ideation to *express thoughts* or *exchange ideas*. Ferguson refers to such sketches as "thinking" and "talking" sketches, respectively.[84]

If the operational principles of artifacts are indeed formulated in the technological mind and communicated as pictures, then it is not unreasonable to believe that one's knowledge of such principles are also held in visual terms, and that the technologist may invoke such visual knowledge during the design process. Such a possibility is illustrated by David Billington's brilliant study of the Swiss bridge designer Robert Maillart (1872–1940), who between 1896 and 1940 designed some forty-seven bridges, mostly of reinforced concrete.[85] Billington showed through his detailed case studies of Maillart's bridges that Maillart chose or conceived the structural form for a bridge such that its behavior under loading—the forces generated within the structure—*could be readily visualized* from the form, thereby greatly simplifying the analysis or calculation of the internal forces. For Maillart, Billington wrote, *force followed form*.[86]

Another example of the way in which operational-principles-as-pictures are stored and employed as knowledge is given by David Pye. In *The Nature and Aesthetics of Design*, Pye describes and analyzes the process he followed in designing a lever system to raise a horizontal beam standing on two columns under the two ends of the beam. Pye's cogitation appears to be strongly visual, both by way of a mental pictorial model of the problem situation and through the use of thinking sketches. And, after completing the design of a rather unusual lever, Pye puzzles over the knowledge source of his idea:

> Where then did the idea of the doubling round the pivot come from? There is a certain element of invention or creation in the idea. . . .
> . . . But I do not believe that this or any such ingenuity is unprompted. Invention involves an antecedent of some kind.
> In the present instance, when I set out to trace in my unconscious mind the antecedent for the doubled lever I found that *there was persistently in my mind's eye* a picture of something not only doubled but also crooked. . . . I realised that this was probably a recollection of an oar called a Yuloh used in China for sculling Junks and Sampans by the stern. [italics added][87]

From his account, it seems reasonable to assert that Pye's personal knowledge body contained a token corresponding to the oar type called the Yuloh. And that, furthermore, this token was represented in a pictorial or diagrammatic form.

10

A Portrait of the Technologist
as a Creative Being

Any explanation of the nature of the technological process that lays even the most modest of claims to generality must necessarily reflect and recognize a multitude of dimensions. It must take into account that technology is born of society—that it is a "social construct;" that any particular episode of technological innovation and change is, frequently, dependent on the personalities, visions, and idiosyncrasies of its principal protagonists; that the choices and decisions that determine the pathways of technological history are contingent upon the structures of power and patronage that prevail in organizations and institutions; and that technical change is inextricably tangled amid the web of economics, commerce, and the profit motive.

These observations apply as much to the matter of creativity as to other facets of technology. Yet ultimately, stripped to its essence, technological creativity—the invention and design of significantly original artifacts of varying degrees of complexity—is an intellectual, cognitive act. Naturally, the intellect so concerned is itself heir to and the product of the influences just noted. It is also shaped by a myriad of personal and social values, knowledge, education, personality traits, and a host of other factors.

It is precisely this complex, many-chambered backdrop that poses the most formidable challenge to those interested in technological creativity. For the question it raises is whether, despite this rich backdrop, it is possible to provide an empirically justified and testable account of how human beings conceive and bring to fruition artifacts in response to practical problems engendered by need, dissatisfaction, or, sometimes, simple curiosity.

As stated in the Prolegomenon, we can call such an account, a *theory* of technological creativity. An attempt to identify the significant elements of such a theory using a marriage of history and cognitive ideas as the tools for understanding is what this book has been about. Let us now see the composite portrait of the creative technologist that this account has unveiled.

The Technological Process Is Knowledge Rich

To begin with, the technologist, viewed as an intellectual cognitive being, operates in a world of goals, action, knowledge, and intendedly rational behavior.[1]

Allen Newell named this cognitive universe the knowledge level. Technological problems may originate in need, dissatisfaction and curiosity—that is, in what philosophers of mind call certain intentional states. But such problems are translated into goals. And the characteristic substance of a technological goal is to produce an artifact that will satisfy certain desired properties and constraints—these being called the requirements for the artifact. Technological goals, then, initiate or drive the process of design and invention.

The creative technologist, contrary to what popular myth proclaims, is armed with a rich body of interconnected, dynamically changing knowledge. Some of this knowledge is not specific to technology at all, but is common to all thinking beings, although the technologist may employ such knowledge more frequently and routinely than others. Examples are the very general mental tools for problem solving, including the variety of rules that allow one to infer something from other things, or the heuristic strategy of means-ends analysis. Knowledge that is specific to the technologist as such—*technological knowledge*—is itself heterogenous in nature. For the technologists of the past two centuries, it includes the basic sciences, mathematics and the body of formal engineering knowledge generally called technological theory (or engineering science). But above all, the knowledge that most distinguishes the technologist from other knowing beings and that has been common to all creators of artifacts from the dawn of humankind to the age of space and the computer is the knowledge of operational principles. The essence of what makes a person a technologist is that his cognitive acts of creation are driven basically by his knowledge of operational principles. This, of course, is a paraphrase of the *operational-principles hypothesis*.[2]

Knowledge may well provide the basis for the technologist to invent and design, but knowledge for its own sake is not what is significant to her. The role of knowledge is to enable the technologist to embark upon cognitive tasks that will further the achievement of technological goals. Just as operational principles distinguish the technologist from other knowing beings, so also the specific nature of these cognitive acts separate out the technologist from other thinking or acting beings. This special kind of cognitive act is what is usually called *design*.

So how does the technologist use knowledge to achieve technological goals? She does so by searching her body of knowledge for, and retrieving, knowledge "tokens"—facts, rules, hypotheses, concepts, beliefs, and so on—that appear relevant to the goal at hand and, on the basis of such knowledge, performing *actions*. Retrieval of knowledge occurs by the activation of tokens linked with the goal by association, and the subsequent associative spread of such activation through the knowledge body.[3]

In the task of designing, the actions performed are mental or "knowledge-level" actions. Such actions transform symbolic representations of physical or abstract things into other symbolic representations. These representations (of goals, the environment, artifacts, or relationships between entities) exist within the technologist's mind, but they can be externalized in the form of text, sketches, mathematical equations, or engineering drawings.

The Technologist Is a Rational Being

In choosing and performing actions in response to goals, the technologist aspires to rational behavior. That is, if he knows that the performance of a certain mental action—a generalization or a logical deduction, for example—will lead to the attainment of a certain goal, then he will choose to perform that action. This is the behavioral *principle of rationality*.[4] Actions so performed are "rational" actions. However, it is quite conceivable for the technologist to execute an action in the absence of a goal. Such action may be called "nonrational."

But even in the case of rational actions, appearances can be quite deceptive. Suppose we observe (by whatever conceivable means) a technologist performing a certain action in response to a certain goal. We infer that he is behaving according to the rationality principle—we think that he "knows what he is doing." Our inference may be quite unfounded, however, for the knowledge on the basis of which the action was taken may be wrong. Or it may be incomplete. He may know the truth but not the whole truth. Or worse, the "space" of possible choices of actions may be so large, as it generally is in the realm of technological problem solving, that the technologist may have neither the cognitive capability nor the luxury of sufficient time to determine the correct action to perform. Though aspiring to behave according to the ideal of "perfect" rationality, the technologist is constrained by, and works under, conditions of *bounded rationality*.[5]

Now, any organized set of actions, whether mental or physical, constitutes a *process*. Thus, in enacting an organized set of cognitive actions, the technologist performs a cognitive (knowledge-level) process. If the first action in this process is invoked in response to a goal, the process as a whole may be said to be a *rational process*. This leads us to the following picture of the technologist as designer: The design process is a rational knowledge-level process (a) initiated or stimulated by a goal to produce an artifact that must satisfy a set of requirements, and (b) terminating in an artifactual form, the design, such that if an artifact is made according to this form, the goal will be achieved.

Ideation Is a Rational Process

The portrait drawn so far is that of the technologist as a cognitive being. But not all acts of design are necessarily creative in the most elevated sense of this word. The most creative kind of the design act is *invention*. This produces artifactual forms that are judged to be significantly original at the very least by the inventor-designer, but more acceptably, by the relevant community. That is, "true" invention engenders artifactual forms that are psychologically or (better still) historically original.[6]

How is the portrait of the technologist as a cognitive being affected by that person being truly inventive? To begin with, recall that all acts of creation involve, at the very least, certain momentous cognitive events called "ideation."[7] Ideation in technology is exemplified by such historical episodes as when Robert Stephenson in 1844 conceived the tubular form for the bridge to span the Menai Strait; when in the 1930s metallurgists at the Mond Nickel Company came up with the idea of employing age-hardened nickel-chromium alloys as the material for gas-turbine blades; when in the 1760s James Watt was first struck with the idea of a separate condenser as an improvement for the Newcomen engine; or when in 1950 Maurice Wilkes thought of the idea of microprogramming for improving the design of computer control units.

Generally speaking, an invention in technology or a discovery in science, a literary product, or a work of art can entail one or many ideations. Whatever may be the case, if we ask what happens in ideation, the answer is, *nothing more than a knowledge-level process*! That is, if for some technological goal, a solution is produced—an artifactual form—that is judged to be historically original (and, thus, a true invention), then the cognitive process performed by the technologist can be described as a knowledge-level process for which the goal is the input and the artifactual form the output.

This, of course, is the *knowledge-level hypothesis on ideation*.[8] According to this hypothesis, the act of invention, of ideation in general, is qualitatively no different from other more mundane instances of the knowledge-level process. However, if we are to follow how ideation in technology may come about, by examining specific historical cases, the following important ramifications can be observed:[9]

First, inventing, being goal directed, is a *purposive* endeavor; invention comes about because the technologist seeks to satisfy a particular practical, technological goal.

Second, creative acts in technology are highly *opportunistic* in nature. They rely on the spawning of subgoals that seem appropriate (according to the rationality principle) to further the achievement of the parent goal but (because of bounded rationality) are not guaranteed to do so. They merely serve as working hypotheses for further exploration.

Third, acts of creation are *gradualistic* in nature. The large insights that ideation produces are composed of a network of small steps. The achievement of a major goal is effected by a network of subgoals, sub-subgoals, and so on, each of which is more tractable and modest in scope than the goal it was spawned from.

Fourth, creative acts of technology are *reasoning* processes, involving the application of rules of inferences. "New" ideas or tokens of knowledge, whether they are facts, rules, hypotheses, or "new" goals, are the consequences of inferences based on generalization, particularization, abduction, or logical deduction applied to combinations of old tokens of knowledge and/or goals. In particular, bisociation—the phenomenon in which two or more seemingly unrelated ideas, goals, or knowledge items combine to form a new idea—is as much explicable in terms of such modes of reasoning as are less startling acts of creation.

Fifth, technological creation is *knowledge intensive*, where the relevant knowledge includes basic science, mathematics, technological theory, commonsense heuristics and problem-solving strategies, and, above all, operational principles.

Finally, the inventor is a cognitive being who *searches freely and associatively* about his or her knowledge body and, being opportunistic, retrieves and

applies whatever knowledge tokens that appear relevant to the goal of the
moment.

The Technologist Freely Espouses and Forms Hypotheses

Creativity in the realm of invention and design entails more than arriving at a
significantly original idea. Stephenson's concept of the tubular form was just
the beginning of the process that led eventually to what came to be known and
admired as the Britannia Bridge. While the tubular form was undoubtedly an
original idea, it would probably have not entered the annals of technological
history of its own. It was the eventual completed, tested, and demonstrably
realizable artifactual form that was acknowledged for its historical originality
rather than the tubular idea alone. In general, an idea must be transformed into
a mature, visible, implementable form to count as an invention.

There is, then, an aspect to technological creativity associated with the
maturation of an idea. Maturation may possibly consume a substantial amount
of time or it may not: The development of the Britannia Bridge consumed three
years, the invention of the first high-temperature superalloys took two, while
the experimental subject who constructed an algorithm for finding the convex
hull did so in a quarter of an hour.[10] Thus, the length of time itself is not a characteristic of the maturation process. But the fact of maturation suggests that
technological creativity is a *sustained* activity.

The production of a mature form is itself performed gradually, by way of a
knowledge-level process. However, the presence of bounded rationality imposes
a definite structure on the process as a whole. More precisely, the creative technologist proceeds from an initial goal through a succession of stages to an "adult"
form. Each stage is characterized by the formation of one or more hypotheses.
The transitions between stages are effected by the critical testing of hypotheses
and their modification into new or revised hypotheses. This process is described
succinctly by the *hypothesis law of maturation*.[11] The creative technologist, thus,
is as much concerned with the espousal of testable hypotheses as the natural scientist. However, the former may not express them explicitly *as* hypotheses.

The process of maturation and development of an artifactual form can be
likened to the biological concept of ontogeny, a word that refers to the "life
history of an individual, both embryonic and postnatal." The hypothesis law
of maturation, then, describes how the *ontogenic development* of artifactual
form is conducted.

Technological Creativity Is Conditioned
by Evolutionary History

We have already seen that ideation is a knowledge-intensive enterprise. One
particular kind of knowledge that pervades technology is that of past, even
vanished, artifacts. Maurice Wilkes drew upon his knowledge of the MIT

Whirlwind computer in deriving his principle of the microprogrammed control unit. Stephenson's tubular bridge in its early form had vestiges of both the suspension and the truss bridges. John Mauchly and Presper Eckert, the designers of the ENIAC were strongly influenced by such prior machines as the mechanical differential analyzer, IBM electromechanical calculators, and John Atanasoff's digital computer. Thomas Newcomen's atmospheric engine bore traces of prior devices and techniques, developed by others, that had exploited the properties of steam, vacuum, and air.

Thus, history is present in one way or another in acts of technological creation. Artifacts, even the most innovative kind, possess evolutionary pasts.

The nature of this evolutionary history is not ontogenic. It is a longer-term phenomenon, involving years, decades, even in some cases, centuries. It entails a historically linked network of mature artifactual forms, the end product of which is the newly invented form. This phenomenon was termed *phylogenic history*, since *phylogeny* in biology refers to the evolutionary history of a lineage, pictured as a series of adult stages.

In sum, technological creativity is conditioned by the evolutionary past. Every act of invention or design has a phylogenic history. This, of course, is the *phylogenic law*.[12]

Epilogue

Creativity is manifested in virtually every significant aspect of the human experience, among which technology stands out for a number of reasons.

First, because it is intimately woven into the fabric of human existence across so many dimensions—personal, societal, economic, national, and international—it provokes strong emotions—exhilaration, admiration, wonder, distaste, fear, and loathing depending on one's view. Second, unlike many other domains of the creative spirit, but in common with the scientific endeavor, its fruits are exposed to the fiercest and hardiest tests of survivability: only those artifactual forms that withstand the storm and stress of the real world are acknowledged and accepted as acts of true invention. But whereas science deals to some extent with idealized and approximate models of the real world, technology (like organisms) must cope with raw nature. Scientific laws at best approximate nature's behavior. But artifacts are far from idealized forms. There is something dramatically substantial about steam engines, cathedrals, skyscrapers, computers, aircrafts, automobiles, bridges, and blast furnaces. A scientific theory that fails will at best be revised or modified and at worst be consigned to the scrapheap of history. If an artifact fails, its consequences are felt in tangible, poignant, and, sometimes, terrible ways. As the failures of nuclear power plants, chemical plants, aircraft, bridges, and pharmaceutical products have shown, consequences reverberate within and across nations. Unlike the artist, writer, or scientist, the technologist must be creative in a *responsible* way.

Third, though now firmly paired with science, technology is as old as mankind, whereas science, or what was formerly called natural philosophy, is scarcely five hundred years old. Thus, invention, design, and the making of artifacts is fundamentally independent of science and, indeed, as independent a human endeavor as are the practices of art, science, and literature.

Because of all these reasons, technology as a creative activity demands and deserves its own attention. In the recent past, such attention has been bestowed on it primarily by biographers and historians. Samuel Smiles's celebrated multivolume study, *Lives of the Engineers*, and his lesser-known *Industrial Biography: Iron Workers and Tool Makers* are pioneering examples of this genre from nineteenth-century England, while in our own century, historically minded engineers and professional historians of technology have provided invaluable records of the circumstances under which specific technological achievements have been effected.

During the past decade or so, a new breed of investigators, armed with paradigms and analytical tools derived from cognitive psychology on the one hand and computer science on the other, have begun to probe into the cognitive processes of discovery and invention. Most of these studies have dealt with specific historical cases: The discovery of the ornithine cycle of urea synthesis; the identification of quantitative and qualitative physicochemical laws such as Kepler's, Boyle's, and Black's laws; the invention of the wave theory of sound; the mechanism by which conceptual revolutions and paradigm shifts in science occur; and the invention of the telephone in the last century and the microprogrammed control unit in this.

In contrast to these earlier studies, I have concerned myself in this book with a systematic inquiry into the creative aspects of technology *as a whole*, using, as supporting evidence, a large variety of examples from different domains and historical times. The outcome of this particular inquiry is intended to be the first step toward a general cognitive theory of technological creativity. If this book prompts the reader to look at technology with new eyes or, still better, manages to stimulate others to probe further into the psychological complexities of technology, then I will feel that it has not been written entirely in vain.

Notes

Chapter 1

1. S. Chandrasekhar, *Truth and Beauty*, University of Chicago Press (Chicago, 1987), p. 2.

2. H. A. Simon, *The Sciences of the Artificial*, 2nd edition, MIT Press (Cambridge, Mass., 1981).

3. S. Dasgupta, *Design Theory and Computer Science*, Cambridge University Press (Cambridge, U.K.,1991).

4. J. C. Jones and D. G. Thornley (eds.), *Conference on Design Methods*, Pergamon Press and Macmillan (Oxford and New York, 1963); C. Alexander, *Notes on the Synthesis of Form*, Harvard University Press (Cambridge, Mass., 1964); D. Pye, *The Nature of Design*, Studio Vista and Rheinhold Publishing Corporation (London and New York, 1964). Pye's book has been republished in revised form as *The Nature and Aesthetics of Design*, Herbert Press (London, 1978).

5. D. E. Knuth, *The Art of Computer Programming*, Addison-Wesley (Reading, Mass.), *Volume 1: Fundamental Algorithms* (1968; 2nd edition 1973); *Volume 2: Seminumerical Algorithms* (1969; 2nd edition, 1981); *Volume 3: Sorting and Searching* (1973). The idea of programming as art is discussed by Knuth at some length in his ACM Turing Award Lecture, "Computer Programming as an Art," *Communications of the ACM* 17 (Dec. 1974), pp. 667–673. This lecture is reprinted in D. E. Knuth, *Literate Programming*, Center for the Study of Language and Information (Stanford, Calif., 1992), pp. 1–16.

6. H. Butterfield, *The Origins of Modern Science: 1300–1800*, Clarke, Irwin and Co. (Toronto, 1968).

7. The essential independence of technology from science has been elegantly argued with the support of many examples from the history of technology by George Basalla in his *The Evolution of Technology* (Cambridge University Press [Cambridge, 1948]). For the historian of technology, a highly influential paper on this topic is E. Layton, "Technology as Knowledge," *Technology and Culture*, 15 (Jan. 1974), 31–41. The argument has also been made recently, and rather forcefully, by Lewis Wolpert in *The Unnatural Nature of Science* (Faber and Faber [London, 1992]).

8. The classic discussion of the origins of the science of strength of materials is S. Timoshenko, *History of Strength of Materials*, McGraw-Hill (New York, 1953; reprint Dover [New York, 1983]). Galileo's contribution to the subject is contained in his *Dialogue Concerning Two New Sciences* (1638), translated by H. Crew and A. deSalvio, Dover (New York, 1954). For a recent discussion of the history of the theory of structures, see W. Addis, *Structural Engineering: The Nature of Theory and Design*, Ellis-Horwood (Chichester, U.K., 1990).

9. T. S. Kuhn, *The Structure of Scientific Revolutions*, 2nd edition, University of Chicago Press (Chicago, 1970; 1st edition 1962).

10. D. A. Norman, *The Design of Everyday Things*, Doubleday (New York, 1989).

11. In *Beyond a Boundary* (Stanley Paul [London, 1963]), C. L. R. James describes how life and the sport of cricket are inextricably bound together in the West Indies. Cricket was not "just a game" on the Caribbean Islands. To know West Indian cricket was to know West Indian life. Or, as James put it, "What do they know of cricket who only cricket know?"

12. For discussions of knowledge-based design, see, e.g., R. D. Coyne, et al., *Knowledge Based Design Systems*, Addison-Wesley (Reading, Mass., 1989), M. D. Rychener (ed.), *Expert Systems for Engineering Design*, Academic Press (New York, 1988), and D. C. Brown and B. Chandrasekaran, *Design Problem Solving*, Pitman (London, 1989).

Chapter 2

1. G. Sturt, *The Wheelwright's Shop*, Cambridge University Press (Cambridge, 1923; Canto edition, 1993).

2. *The Oxford English Dictionary*, 2nd edition, Oxford University Press (Oxford, 1990).

3. G. Basalla, *The Evolution of Technology*, Cambridge University Press (Cambridge, 1988).

4. H. A. Simon, *The Sciences of the Artificial*, 2nd edition, MIT Press (Cambridge, Mass., 1981; 1st edition 1969).

5. H. Petroski, *The Pencil: A History of Design and Circumstance*, Alfred A. Knopf (New York, 1989).

6. K. R. Popper, *Objective Knowledge*, Clarendon Press (Oxford, 1972).

7. J. C. Jones, *Design Methods: Seeds of Human Future*, John Wiley & Sons (New York, 1980).

8. C. Alexander, *Notes on the Synthesis of Form*, Harvard University Press (Cambridge, Mass., 1964).

9. B. Ghiselin (ed.), *The Creative Process*, University of California Press (Berkeley, 1952).

10. E. Clark, *The Britannia and Conway Tubular Bridges*, 2 volumes, Day & Sons (London, 1850); W. Fairbairn, *An Account of the Construction of the Britannia and Conway Tubular Bridges*, John Wheale/Longman, Brown, Green and Longmans (London, 1849). An important recent modern account of the Britannia Bridge project is N. Rosenberg and W. G. Vincenti, *The Britannia Bridge: The Generation and Diffusion of Technological Knowledge*, MIT Press (Cambridge, Mass., 1978).

11. See. e.g., Petroski, op. cit., pp. 8–10; E. Ferguson, *Engineering and the Mind's Eye*, MIT Press (Cambridge, Mass., 1992).

12. See the Foreword by E. P. Thompson to the Canto edition of Sturt, op. cit.

13. Sturt, op. cit., p. 31.

14. Sturt, op. cit., pp. 19–20.

15. R. C. Atkinson and R. M. Shiffrin, "Human Memory: A Proposed System and Its Control Processes," in *The Psychology of Learning and Motivation*, Vol. 2, K. Spence & J. Spence (eds.), Academic Press (New York, 1968); G. A. Miller, "The Magical Number Seven, Plus or Minus Two: Some Limits on Our Capacity for Information Processing," *Psychological Review*, 63 (1956), 81–97; R. Lachman, J. L. Lachman, and

E. C. Butterfield, *Cognitive Psychology and Information Processing*, Lawrence Erlbaum and Associates (Hillsdale, N.J., 1979).

16. Vitruvius, *De Architectura* (c. 25 B.C.). Translated by M. H. Morgan as *Ten Books on Architecture*, Harvard University Press (Cambridge, Mass., 1914; reprint, Dover [New York, 1960]).

17. W. Addis, *Structural Engineering: The Nature of Theory and Design*, Ellis Horwood (Chichester, U.K., 1990), pp. 115–122.

18. Addis, op. cit., p. 123.

19. S. Smiles, *Industrial Biography: Iron Workers and Tool Makers*, John Murray (London, 1863; reprinted with Introduction by L. T. C. Rolt, David & Charles [Newton Abbot, U.K., 1967]), pp. 86–88.

20. Smiles, op. cit., pp. 87–88.

21. Smiles, op. cit., p. 88.

22. The reverberatory furnace was invented in 1613 by one Rovenson (first name not known) who described it in his *Treatise of Metallica* published that same year. In such a furnace, the material to be melted or to be wrought is kept separate from direct contact with the fuel and is heated by the air, flame, or gases from the fuel being reflected or "reverberated" down from the furnace roof onto the material.

23. For a personal retrospective account of the development of the EDSAC, see M. V. Wilkes, *Memoirs of a Computer Pioneer*, MIT Press (Cambridge, Mass., 1985). A more technical discussion is given in M. V. Wilkes, *Automatic Digital Computers*, Methuen & Co. (London, 1957).

24. C. Babbage, *The Works of Charles Babbage*, M. Campbell-Kelly (ed.), William Pickering (London, 1989), 9 volumes. The Analytical Engine is described in Volume 3. See also C. Babbage, *Passages from the Life of a Philosopher*, M. Campbell-Kelly (ed.), Pickering & Chatto (London, 1992), and M. V. Wilkes, "Babbage as a Computer Pioneer," *Historia Mathematica*, 4 (1977), 415–440.

25. The view of programs or software as mathematical entities originated in two seminal papers published in the late 1960s: R. W. Floyd, "Assigning Meaning to Programs," *Mathematical Aspects of Computer Science*, American Mathematical Society (Providence, R.I., 1967); and C. A. R. Hoare, "An Axiomatic Approach to Computer Programming," *Communications of the ACM*, 12, no. 10 (Oct. 1969), 576–580, 583. A more recent and quite accessible discussion of the mathematical nature of programs is C. A. R. Hoare, "The Mathematics of Programming," Inaugural Lecture, Univ. of Oxford, Clarendon Press (Oxford, 1986). For a discussion of software as abstract artifact, see S. Dasgupta, *Design Theory and Computer Science*, Cambridge University Press (Cambridge, 1991), especially Chapter 8.

26. For a discussion of the causal or physical nature of programs, see J. Fetzer, "Program Verification: The Very Idea," *Communications of the ACM*, 31, no. 9 (Sept. 1988), 1048–1063. A summary of Fetzer's argument is given in Dasgupta, op. cit., 228–232.

27. See, e.g., D. Gries, *The Science of Programming*, Springer-Verlag (New York, 1981); E. W. Dijkstra, *A Discipline of Programming*, Prentice-Hall (Englewood-Cliffs, N.J., 1976).

28. See, e.g., I. Sommerville, *Software Engineering*, 3rd edition, Addison-Wesley (Reading, Mass., 1989).

29. See also the Prolegomenon.

30. D. Pye, *The Nature and Aesthetics of Design*, Herbert Press (London, 1978), p. 13.

31. P. B. Medawar, *The Art of the Soluble*, Penguin Books (Harmondsworth, U.K., 1969), p. 97. Derek Gjertsen, however, has argued that scientists, in addition to solving problems, engage in a variety of other kinds of activities: They measure, for example, and classify and catalogue and so on. Gjertsen has also pointed out, using historical examples, that scientists (including mathematicians) are often concerned with longstanding unsolved problems—as, for example, finding a proof for Fermat's famous "last theorem." See D. Gjertsen, *Science and Philosophy*, Penguin Books (London, 1989), Chapter 3.

32. Rosenberg and Vincenti, op. cit., p. 43.

33. H. Petroski, *Design Paradigms: Case Histories of Error and Judgement in Engineering*, Cambridge University Press (New York, 1994). See, especially, Chapter 7, "The Britannia Tubular Bridge: A Paradigm of Tunnel Vision in Design," pp. 99–120.

34. Clark, op. cit., pp. 83–205, Fairbairn, op. cit., 209–283; also, Rosenberg and Vincenti, op. cit.

35. J. Reason, *Human Error*, Cambridge University Press (Cambridge, 1990).

36. Reason, op. cit., p. 9.

Chapter 3

1. E. H. Gombrich, *Art and Illusion: A Study in the Psychology of Pictorial Representation*, Phaidon Books (London, 1960).

2. *The Concise Oxford English Dictionary*, 8th edition, R. E. Allen (ed.), Clarendon Press (Oxford, 1990).

3. *Creep* is the phenomenon of continuous, time-dependent plastic (i.e., nonreversible) deformation of metals and alloys under load. In general, metals deform more easily under load at elevated temperatures then at room temperature. Thus alloys that are expected to operate at high temperatures (e.g., those used in the construction of gas-turbine blades) are more likely to exhibit creep than materials operating at lower, say room, temperatures. For more on creep, see, e.g., H. E. McGannon (ed.) *The Making, Shaping and Treating of Steel*, 9th edition, United States Steel (Pittsburgh, Penn., 1970).

4. R. W. Cahn, "Modern Practice in the Design of Strong Alloys," *Recent Developments in Metallurgical Science and Technology*, Vol. 4, Indian Institute of Metals, Silver Jubilee Symposium, (New Delhi, 1972) pp. 527–551.

5. "Large" here is, of course, relative to the time we are talking about. In the mid-1940s, prior to the actual development of stored-program digital computers, when the ENIAC represented the primary electronic exemplar of digital computers, a memory capable of storing a thousand or so "bits" of information was considered large!

6. A. W. Burks, "From ENIAC to the Stored-Program Computer: Two Revolutions in Computers," in N. Metropolis, J. Howlett, and G.-C. Rota (eds.), *A History of Computing in the Twentieth Century*, Academic Press (New York, 1980), pp. 311–344.

7. S. Smiles, *Industrial Biography: Iron Workers and Tool Makers*, John Murray (London, 1863; reprinted with Introduction by L. T. C. Rolt, David & Charles [Newton Abbot, U.K., 1967]), p. 48. In his Introduction to the 1967 reprint of the 1863 edition, L. T. C. Rolt makes the point that Smiles credited Dud Dudley with the invention of coal-based iron smelting on the basis of Dudley's own account in *Metallum Martis*. However, Rolt points out that according to recent historical researches, Dudley could not have produced a successful process and that the credit for this process must, properly, go to Abraham Darby I, the founder of the famous Coalbrookdale Works in Shropshire in 1704. (Smiles, op. cit., pp. 80–83).

8. Smiles, op. cit., pp. 40–41, pp. 43–44.

9. E. Clark, *The Britannia and Conway Tubular Bridges*, 2 volumes, Day & Sons (London, 1850); N. Rosenberg and W. G. Vincenti, *The Britannia Bridge: The Generation and Diffusion of Technological Knowledge*, MIT Press (Cambridge, Mass., 1978).

10. Burks, op. cit. See also J. P. Eckert, "The ENIAC," Metropolis, Howlett, and Rota, op. cit., pp. 525–539; J. W. Mauchly, "The ENIAC," Metropolis, Howlett, and Rota, op. cit., pp. 541–550.

11. Burks, op. cit., p. 334.

12. Eckert, op. cit., p. 326.

13. Burks, op. cit., p. 314.

14. W. Aspray, *John von Neumann and the Origin of Modern Computing*, MIT Press (Cambridge, Mass., 1990), p. 36.

15. Aspray, op. cit., p. 36.

16. Burks, op. cit., pp. 327–333, gives an excellent description of how an ENIAC program would be manually set up.

17. The EDSAC successfully executed its first program in the middle of 1949 (M. V. Wilkes, *Memoirs of a Computer Pioneer*, MIT Press, Cambridge, Mass., 1985, p. 142). The MARK I, developed by F. C. Williams, T. Kilburn, and their collaborators in the University of Manchester, actually began to execute programs at about the same time as the EDSAC, but it was somewhat later that the MARK I had regular input and output devices. An earlier version of the Manchester machine, known as the "baby MARK I," was actually demonstrated on June 21, 1948, but this, apparently, had only rudimentary arithmetic capabilities (M. V. Wilkes, personal communications, January 29, 1986). For a discussion of the history of the MARK I, see S. H. Lavington, "Computer Developments at Manchester University," in Metropolis, Howlett, and Rota, op. cit., pp. 433–443.

18. The original paper on microprogramming is M. V. Wilkes, "The Best Way to Design an Automatic Calculating Machine," *Report, Manchester University Computer Inaugural Conference*, Manchester, U.K., July, 1951. This has been reprinted in *Annals of the History of Computing*, 8, no. 2 (Apr. 1986), 118–121. For retrospective accounts of the circumstances attending this invention, see M. V. Wilkes, "The Genesis of Microprogramming," *Annals of the Hist. of Computing*, 8, no. 2 (Apr. 1986), 116–118; and M. V. Wilkes, *Memoirs of a Computer Pioneer*, MIT Press (Cambridge, Mass., 1985), pp. 184–185. A detailed cognitive explanation of Wilkes's invention of microprogramming, upon which the present discussion and those later in this book are based, is given in S. Dasgupta, *Creativity in Invention and Design*, Cambridge University Press (New York, 1994).

19. See Dasgupta op. cit., pp. 14–15.

20. M. V. Wilkes, "The Origin and Development of Microprogramming," Inauguration Videotaped Lecture, International Repository on Microprogramming, Dupré Library, University of Southwestern Louisiana, Lafayette, Louisiana, Oct. 29, 1984.

21. W. B. Carlson and M. E. Gorman, "A Cognitive Framework to Understand Technological Creativity: Bell, Edison and the Telephone," in R. J. Weber and D. N. Perkins (eds.), *Inventive Minds: Creativity in Technology*, Oxford University Press (New York, 1992), pp. 48–79.

22. Carlson and Gorman, op. cit., p. 54.

23. Carlson and Gorman, op. cit., p. 54.

24. For a history of the development of the crucible process, see K. C. Barraclough, *Steelmaking before Bessemer*, Vol. 2 *Crucible Steel*, Metals Society (London, 1984).

For biographical details of Benjamin Huntsman, see Smiles, op. cit., pp. 99–113; K. C. Barraclough, "Benjamin Huntsman, 1704–1776," Sheffield City Libraries, Local Studies Booklet, 1976; E. W. Hulme, "The Pedigree and Career of Benjamin Huntsman, Inventor in Europe of Crucible Steel," *Trans. Newcomen Society*, 24 (1944/45), 37–48.

25. Barraclough, in his booklet "Benjamin Huntsman, 1704–1776," op. cit., p. 2, has pointed out that the term *German steel* is, perhaps, confusing. Steel made in Austria or Westphalia and imported via the Rhine was indeed called "German steel." But so was the product made at the same time at the Blackhall Mill on the River Derwent—this later coming to be known as "shear steel."

26. Barraclough, op. cit., p. 1; Smiles, op. cit., p. 102. Both Barraclough and Smiles cite Professor Le Play, professor of metallurgy in the Royal School of Mines in France who, after examining at length the Sheffield steel industry in the late 1830s and early 1840s firmly credited Huntsman as the prime mover in the advancement of the British steel industry between the mid-eighteenth and mid-nineteenth centuries.

27. W. Addis, *Structural Engineering: The Nature of Theory and Design*, Ellis Horwood (Chichester, U.K., 1990), p. 185.

28. Addis, op. cit., pp. 97–112.

29. T. S. Kuhn, *The Structure of Scientific Revolutions*, 2nd edition, University of Chicago Press (Chicago, 1970; 1st edition 1962).

30. Addis, op. cit., pp. 102–103.

31. J. F. Baker, *The Steel Skeleton: Elastic Behavior and Design*, Cambridge University Press (Cambridge, 1954), p. 6.

32. M. V. Wilkes, in an interview with the author, Dec. 19, 1991, Olivetti Research Laboratory, Cambridge.

33. R. K. Jurgen, "Jacob Rabinow," *IEEE Spectrum*, Dec. 1991, pp. 24–25.

34. Quoted in Jurgen, op. cit., p. 24.

35. Quoted in G. F. Watson, "Masaru Ibuka," *IEEE Spectrum*, Dec. 1991, pp. 22–28.

36. D. E. Knuth, *Computers and Typesetting*, 5 volumes, Addison-Wesley (Reading, Mass., 1986).

37. D. E. Knuth, "The Errors of TEX," *Software: Practice and Experience*, 19, no. 7 (1989), 607–685. Reprinted in D. E. Knuth, *Literate Programming*, Center for Language and Information, Stanford University (Stanford, Calif., 1992).

38. D. E. Knuth, undated personal communication to the author, letter received June 23, 1993.

39. J. Searle, *Intentionality: An Essay in the Philosophy of Mind*, Cambridge Univ. Press (Cambridge, 1983), D. C. Dennett, *The Intentional Stance*, MIT Press (Cambridge, Mass., 1987); J. Searle, *The Rediscovery of Mind*, MIT Press (Cambridge, Mass., 1992).

Chapter 4

1. H. A. Simon, "The Architecture of Complexity," *Proc. Amer. Phil. Soc.*, 106 (Dec. 1962), 467–482. Reprinted in H. A. Simon, *The Sciences of the Artificial*, 2nd edition, MIT Press (Cambridge, Mass., 1981; 1st edition 1969).

2. C. G. Bell and A. Newell, *Computer Structures: Readings and Examples*, McGraw-Hill (New York, 1971), pp. 3–10.

3. See, for example, Z. Pylyshyn, "Computation and Cognition: Issues in the Foundation of Cognitive Science," *Behavioral and Brain Sciences*, 3, no. 1 (1980), 154–169; Z. Pylyshyn, *Computation and Cognition*, MIT Press (Cambridge, Mass., 1984);

A. Newell, *Unified Theories of Cognition*, Harvard University Press (Cambridge, Mass., 1990); A. Newell, P. S. Rosenbloom, and J. Laird, "Symbolic Architectures for Cognition," M. I. Posner (ed.), *Foundations of Cognitive Science*, MIT Press (Cambridge, Mass., 1989), 93–131. See also B. Chandrasekaran and S. G. Josephson, "Architecture of Intelligence: The Problem and Current Approaches to Solutions," *Current Science*, 64, no. 6 (Mar. 25, 1993), for a comprehensive review of the different schools of thought about cognitive levels of description.

4. 1 millisecond = 1/1000th of a second.

5. R. L. Gregory, *The Mind in Science*, Weidenfeld and Nicholson (London, 1981).

6. A. M. Turing, "Computing Machinery and Intelligence," *Mind*, 59 (Oct. 1950), 433–460.

7. For a glimpse (or more) of the technical nature and substance of the discipline of AI, see, for example, A. Barr and E. Feigenbaum (eds.), *Handbook of Artificial Intelligence*, 3 volumes, Morgan Kaufmann (Los Altos, Calif., 1981); E. Charniak and D. McDermott, *Introduction to Artificial Intelligence*, Addison-Wesley (Reading, Mass., 1985); P. Winston, *Artificial Intelligence*, 2nd edition, Addison-Wesley (Reading, Mass., 1984). Important criticisms of AI include H. Dreyfus, *What Computers Can't Do*, revised edition, Harper & Row (New York, 1979); J. R. Searle, *Mind, Brain and Science*, Harvard University Press (Cambridge, Mass., 1984); and, most recently, R. Penrose, *The Emperor's New Mind*, Oxford University Press (New York, 1989). For a broad survey of AI, see also the Winter 1988 issue of *Daedalus*, the Journal of the American Academy of Arts and Sciences.

8. The state of the art of cognitive science is, perhaps, most comprehensively documented in M. Posner (ed.), *Foundations of Cognitive Science*, MIT Press (Cambridge, Mass., 1989). On the relationship between AI and psychology, see M. Boden, *Artificial Intelligence and Psychology*, MIT Press (Cambridge, Mass., 1989). For a less technical account, see P. N. Johnson-Laird, *The Computer and the Mind*, Harvard University Press (Cambridge, Mass., 1988).

9. See, for example, S. Papert, "One AI or Many," *Daedalus*, Winter 1988; also, J. A. Fodor and Z. A. Pylyshyn, "Connectionism and Cognitive Architecture: A Critical Analysis," in S. Pinker and S. Mehler (eds.), *Connections and Symbols*, MIT Press (Cambridge, Mass., 1988), 30–72.

10. Z. Pylyshyn, "Computation and Cognition: Issues in the Foundations of Cognitive Science," *Behavioral and Brain Science*, 3, no. 1 (1980), 154–169; D. Dennett, *Brainstorms*, MIT Press (Cambridge, Mass., 1978). See also D. Dennett, *The Intentional Stance*, MIT Press (Cambridge, Mass., 1988).

11. A. Newell, "The Knowledge Level," *Artificial Intelligence*, 18 (1982), 87–127.

12. M. V. Wilkes, *Memoirs of a Computer Pioneer*, MIT Press (Cambridge, Mass., 1985), pp. 176–182; also, M. V. Wilkes, "The Genesis of Microprogramming," *Annals of the Hist. of Computing*, 8, no. 2 (Apr. 1986), 116–118.

13. The origin of rules of this form actually reaches back to the logician Emil Post's work, in 1943, on the manipulation and transformation of symbolic structures. This same work may have led Noam Chomsky to the idea of characterizing the grammar of a language in terms of "rewriting rules" in the 1950s. The use of rules to represent knowledge in AI and for modeling cognitive behavior appears to be due to Allen Newell in the 1960s. For a discussion of the role of rules in logic and language see, respectively, M. Minsky, *Computation: Finite and Infinite Machines*, Prentice-Hall (Englewood Cliffs, N.J., 1967); and N. Chomsky, *Syntactic Structures*, Mouton (The Hague, 1969). For the use of rules in describing human cognition, see A. Newell and H. A. Simon, *Human Problem Solving*, Prentice-Hall (Englewood Cliffs, N.J., 1972); J. R.

Anderson, *The Architecture of Cognition*, Harvard University Press (Cambridge, Mass., 1983).

14. G. Polya, *How to Solve It*, 2nd edition, Princeton University Press (Princeton, N.J., 1957).

15. See Chapter 9 for a more extensive discussion of technological knowledge.

16. Vitruvius, *Ten Books on Architecture*, translated by M. H. Morgan, Harvard University Press (Cambridge, Mass., 1914; reprint, Dover [New York, 1960]).

17. Vitruvius, op. cit., pp. 110–113.

18. See, e.g., G. Broadbent, *Design in Architecture*, John Wiley & Sons (Chichester, U.K., 1973); H. J. Cowan, *An Historical Outline of the Architectural Sciences*, 2nd edition, Elsevier (New York, 1977); and S. Groak, *The Idea of Building*, E. & F. N. Spon (London, 1992).

19. C. Alexander, S. Ishikawa, and M. Silverstein, *A Pattern Language*, Oxford University Press (New York, 1977).

20. C. Alexander, H. Neis, A. Anninou, and I. King, *A New Theory of Urban Design*, Oxford University Press (New York, 1987), pp. 63–76.

21. G. Agricola, *De Re Metallica* (1556), translated by H. C. Hoover and L. H. Hoover, *The Mining Magazine*, London, 1912. Reprinted by Dover (New York, 1950).

22. A *flux* is any matter added to ores to more easily melt the ore or aid in the separation of the impurities from the metal in one way or another. The fluxes Agricola mentions include red lead or litharge (Pb_3O_4), lead ochre (PbO), and copper.

23. Agricola, op. cit., p. 235.

24. Agricola, op. cit., p. 243.

25. Agricola, ibid.

26. E. W. Dijkstra, "Notes on Structured Programming," in O.-J. Dahl, E. W. Dijkstra, and C. A. R. Hoare, *Structured Programming*, Academic Press (New York, 1972); N. Wirth, "Program Development by Stepwise Refinement," *Communications of the ACM*, 14, no. 4 (Apr. 1971), 221–227.

27. D. E. Knuth, "The Errors of TEX," *Software: Practice and Experience*, 19, July 1989, 607–685. Reprinted in D. E. Knuth, *Literate Programming*, Center for the Study of Language and Information (Stanford, Calif., 1992), 243–292.

28. T. S. Kuhn, *The Copernican Revolution*, Harvard University Press (Cambridge, Mass., 1957), pp. 28–29, 78–94.

29. Kuhn, op. cit., pp. 145–150.

30. For a discussion of Kepler's worldview, see G. Holton, *An Introduction to Concepts and Theories in Physical Sciences*, Addison-Wesley (Reading, Mass., 1952), especially Chapters 2 and 9.

31. Holton, op. cit., especially Chapter 11.

32. For an extensive discussion of Einstein's *Weltanshauung* and its import on his thinking and views on physics, see, for example, A. Pais, '*Subtle is the Lord . . .': The Science and Life of Albert Einstein*, Oxford University Press (Oxford, 1982); also, G. Holton and Y. Elkana (eds.), *Albert Einstein: Historical and Cultural Perspectives*, Princeton University Press (Princeton, N.J., 1982)—especially the article by M. Jammer, "Einstein and Quantum Mechanics," pp. 59–78.

33. R. Sperry, *Science and Moral Priority*, Basil Blackwell (Oxford, 1983).

34. P. L. Nervi, *Aesthetics and Technology in Building*, Harvard University Press (Cambridge, Mass., 1966). See also *The Works of Pier Luigi Nervi*, F. A. Praeger (New York, 1957).

35. D. P. Billington, *Robert Maillart's Bridges: The Art of Engineering*, Princeton University Press (Princeton, N.J., 1979).

36. *Bandwidth* is an appropriate performance measure of any system that transmits or communicates information. It refers to the amount of information transmitted per unit time.

37. B. Chandrasekaran and S. G. Josephson, "Architecture of Intelligence: The Problems and Current Approaches to Solutions," *Current Science*, 64, no. 6 (Mar. 25, 1993), 366–380.

38. A. Newell, "The Knowledge Level," *Artificial Intelligence*, 18 (1982), 87–127. See also his *Unified Theories of Cognition*, Harvard University Press (Cambridge, Mass., 1990).

39. S. Dasgupta, *Creativity in Invention and Design*, Cambridge University Press (New York, 1994).

40. H. Petroski, "Vitruvius's Auger and Galileo's Bones: Paradigms of Limit to Size in Design," *J. Mech. Design (Trans. Amer. Soc: Mech. Engineers)*, 114, (March 1992), 23–28. See, alternatively, Petroski, *Design Paradigms: Case Histories of Error and Judgement in Engineering*, Cambridge University Press (New York, 1994), Chapter 3.

41. W. Reitman, *Cognition and Thought*, John Wiley (New York, 1965); H. A. Simon, "The Structure of Ill-Structured Problems," *Artificial Intelligence*, 4 (1973), 181–200.

42. See D. A. Norman, *The Design of Everyday Things*, Doubleday (New York, 1989), for examples of pedestrian artifacts that are ill-designed.

43. H. A. Simon, *Administrative Behavior*, 3rd edition, Free Press (New York 1976; 1st edition 1946), especially Chapters 2, 4, and 11.

44. Simon's papers and lectures on bounded rationality, including his Nobel lecture, are collectively published as *Models of Bounded Rationality*, 2 volumes, MIT Press (Cambridge, Mass., 1982). A particularly succinct description of the concept of bounded rationality is presented in the paper "Theories of Bounded Rationality," reprinted in Vol. 2, pp. 400–423.

45. See A. Newell, *Unified Theories of Cognition*, Harvard University Press (Cambridge, Mass., 1990); also, D. E. Rumelhart and J. L. McClelland, *Parallel Distributed Processing, Vol. 1: Foundations*, MIT Press (Cambridge, Mass., 1986).

46. Spreading activation was discussed in A. M. Collins and E. F. Loftus, "A Spreading-Activation Theory of Semantic Processing," *Psychological Reviews*, 82 (1975), 407– 428. It also forms one of the cornerstones in J. R. Anderson, *The Architecture of Cognition*, Harvard University Press (Cambridge, Mass., 1983). For more recent discussions, see, e.g., J. H. Holland, K. J. Holyak, R. E. Nisbett, and P. R. Thagard, *Induction*, MIT Press (Cambridge, Mass., 1986); and P. R. Thagard, *Computational Philosophy of Science*, MIT Press (Cambridge, Mass., 1988).

47. Scientists in both AI and cognitive psychology have developed over the years a variety of knowledge-representation schemes. The network constitutes one of these. In the AI literature, such schemes go by several names, including "associative network," "semantic network," and "conceptual network." For more on the topic, see, e.g., R. Brachman and H. Levesque (eds.), *Readings in Knowledge Representation*, Morgan Kaufmann (Los Altos, Calif., 1985); A. Barr and E. Feigenbaum, *Handbook of Artificial Intelligence, Vol. 1*, Morgan Kaufmann (Los Altos, Calif., 1981); and N. V. Findler (ed.), *Associative Network: Representation and Use of Knowledge by Computers*, Academic Press (New York, 1974).

Paul Thagard used network schemes that he called "conceptual networks" as a way of representing chemical and geological knowledge in his discussions of conceptual revolutions in science. See his "The Conceptual Structure of the Chemical Revolution,"

Philosophy of Science, 57 (1990), 183–209 and *Conceptual Revolutions*, Princeton University Press (Princeton, N.J., 1992); See also P. Thagard and G. Nowak, "The Conceptual Structure of the Geological Revolution" in J. Shrager and P. Langley (eds.), *Computational Models of Scientific Discovery and Theory Formation*, Morgan Kaufmann (San Mateo, Calif., 1990).

Conceptual networks of the type proposed by Thagard were also used by this author to describe the invention of microprogramming. See S. Dasgupta, *Creativity in Invention and Design*, Cambridge University Press (New York, 1994).

48. Newell, op. cit., pp. 160 et seq.

49. M. Minsky, "A Framework for Representing Knowledge," in P. Winston (ed.), *The Psychology of Computer Vision*, McGraw-Hill (New York, 1975), pp. 211–277.

50. F. C. Bartlett, *Remembering: A Study in Experimental and Social Psychology*, Cambridge University Press (Cambridge, 1932). See also R. C. Schank and R. P. Abelson, *Scripts, Plans, Goals and Understanding*, Lawrence Erlbaum (Hillsdale, N.J., 1977); and M. A. Arbib and M. B. Hesse, *The Construction of Reality*, Cambridge University Press (Cambridge, 1986).

51. Up to the end of the nineteenth century, it was taken for granted that energy was of a continuous (that is, nondiscrete) nature. However, the German physicist Max Planck showed that the energy emitted by an atom when excited can take on only discrete values.

52. A. Koestler, *The Act of Creation*, Hutchinson & Co. (London, 1964).

53. W. Addis, *Structural Engineering: The Nature of Theory and Design*, Ellis-Horwood (Chichester, U.K., 1990), p. 1.

54. Addis, ibid.

55. Alexander, op. cit., p. 1.

56. D. Pye, *The Nature and Aesthetics of Design*, Herbert Press (London, 1978); S. Dasgupta, *Design Theory and Computer Science*, Cambridge University Press (Cambridge, 1991).

57. Dasgupta, op. cit., p. 6.

58. T. S. Kuhn, *The Structure of Scientific Revolutions*, University of Chicago Press (Chicago 1962; 2nd edition 1970; 1st edition 1962).

59. Newell, op. cit., pp. 102–107.

60. Kuhn, op. cit.

61. W. Vincenti, *What Engineers Know and How They Know It*, Johns Hopkins University Press (Baltimore, Md. 1992).

62. S. Dasgupta, *Creativity in Invention and Design*, Cambridge University Press (New York, 1994).

Chapter 5

1. D. Pye, *The Nature and Aesthetics of Design*, Herbert Press (London, 1978), p. 21.

2. See, e.g., H. E. McGannon (ed.), *The Making, Shaping and Treating of Steel*, 9th edition, U.S. Steel Corp. (Bethlehem, Penn., 1971), p. 24; W. K. V. Gale, "Ferrous Metals," in I. McNeil (ed.), *An Encyclopedia of the History of Technology*, Routledge (London, 1990), pp. 167–168.

3. J. von Neumann, "First Draft of a Report on the EDVAC," Moore School of Electrical Engineering, University of Pennsylvania (Philadelphia, 1945), reprinted in B. Randell (ed.) *The Origins of Digital Computers: Selected Papers*, Springer-Verlag

(New York, 1975), pp. 355–365; see also A. W. Burks, "From ENIAC to the Stored Program Computer: Two Revolutions in Computers" in *A History of Computing in the Twentieth Century*, N. Metropolis, J. Howlett, and G.-C. Rota (eds.), Academic Press (New York, 1980), pp. 311–344.

4. For an account of the early history of microprogramming, see M. V. Wilkes, "The Genesis of Microprogramming," *Annals of the History of Computing*, 8, no. 2 (Apr. 1986), 116–118; see also S. Dasgupta, *Creativity in Invention and Design*, Cambridge University Press (New York, 1994), especially Chapter 3.

5. The dispute over the invention of the electronic digital computer—whether the credit should go the Eckert and Mauchly or to Atanasoff—became a matter for the courts between 1967 and 1973. The outcome of this lawsuit was reported in E. R. Larson, "Findings of Fact, Conclusions of Law and Order for Judgement," File No. 4–67, Civ. 138, *Honeywell, Inc. vs. Sperry Rand Corporation and Illinois Scientific Developments, Inc.*, U.S. District Court, District of Minnesota, Fourth Division (Oct. 19, 1973). For further comments on this see the Appendix in B. Randell, "Origins of Digital Computers: Supplementary Bibliography," in Metropolis, Howlett, and Rota, op. cit., pp. 629–659.

6. G. Basalla, *The Evolution of Technology*, Cambridge University Press (Cambridge, 1988), pp. 35–40; E. F. C. Somerscales, "Steam and Internal Combustion Engines," in McNeil, op. cit., pp. 272–349.

7. W. B. Carlson and M. E. Gorman, "A Cognitive Framework to Understand Technological Creativity: Bell, Edison and the Telephone," in R. J. Weber and D. N. Perkins (eds.), *Inventive Minds*, Oxford University Press (New York, 1992), pp. 48–77.

8. M. Boden, *The Creative Mind*, Basic Books (New York, 1991), p. 32.

9. S. Dasgupta, *Creativity in Invention and Design*, Cambridge University Press (New York, 1994), pp. 15–22.

10. T. S. Kuhn, *The Structure of Scientific Revolutions*, University of Chicago Press (Chicago 1968; 2nd edition 1970; 1st edition 1962). See also P. Thagard, *Conceptual Revolutions*, Princeton University Press (Princeton, N.J., 1992).

11. D. E. Knuth, *Computers and Typesetting*, 5 Volumes, Addison-Wesley (Reading, Mass., 1991).

12. Dasgupta, op. cit., pp. 15–22.

13. Hardy's best-known discussion of Ramanujan's mathematics is his *Ramanujan*, Cambridge University Press (Cambridge, 1940). An earlier and very beautiful account is his 1920 obituary notice on the Indian mathematician published in the *Obituary Notices of the Royal Society*, The Royal Society (London, 1920).

14. R. Kanigel, *The Man Who Knew Infinity*, Charles Scribner's Sons (New York, 1991).

15. C. P. Snow, Forward to the 1967 edition of G. H. Hardy, *A Mathematician's Apology*, Cambridge University Press (Cambridge, 1940), p. 31.

16. Snow, op. cit., p. 32.

17. Snow, op. cit., p. 32.

18. G. H. Hardy, *Ramanujan*, Cambridge University Press (Cambridge, 1940), p. 10.

19. Hardy, op. cit., p. 10.

20. See, e.g., L. Laudan, *Science and Values*, University of California Press (Berkeley, 1984).

21. For a lucid discussion of nonmonotonic reasoning, see R. Reiter, "Nonmonotonic Reasoning," *Annual Review of Computer Science*, 2, 1987, 147–186.

22. For extensive general discussions of scientific revolutions, see, e.g, T. S. Kuhn,

The Copernican Revolution, Harvard University Press (Cambridge, Mass., 1957), and idem, *The Structure of Scientific Revolutions*, University of Chicago Press (Chicago 1962; 2nd edition 1970); I. B. Cohen, *The Newtonian Revolution*, Cambridge University Press (Cambridge, 1980); and idem, *Revolution in Science*, Harvard University Press (Cambridge, Mass., 1985); and, most recently, P. Thagard, *Conceptual Revolutions*, Princeton University Press (Princeton, N.J., 1992).

23. G. Basalla, *The Evolution of Technology*, Cambridge University Press (Cambridge, 1988), pp. 7–8. D. S. L. Cardwell, *The Fontana History of Technology*, Fontana Press (London, 1994), pp. 37–41.

24. W. Addis, *Structural Engineering: The Nature of Theory and Design*, Ellis Horwood (Chichester, U.K., 1990).

25. Addis, op. cit., p. 115.

26. Vitruvius, *De Architectura* (circa 25 B.C.), translated by M. H. Morgan as *Ten Books on Architecture*, Harvard University Press (Cambridge, Mass., 1914; reprint Dover [New York, 1960]).

27. Kanigel, op. cit., pp. 337–340.

28. Kanigel, op. cit., p. 370.

29. G. H. Hardy, *Ramanujan: Twelve Lectures on Subjects Suggested by His Life and Work*, Cambridge University Press (Cambridge, 1940).

30. A. M. Turing, "On Computable Numbers with an Application to the Eintscheidungsproblem," *Proc. London Mathematical Society*, ser. 2., 42 (1936), 230–265.

31. Boden, op. cit., pp. 35–36.

32. This particular episode is mentioned rather offhandedly by Andrew Hodges in his biography of Turing: *Alan Turing: The Engima*, Simon and Schuster (New York, 1983). Hodges records (p. 94) that Turing was elected a fellow in March 1935 on the strength of the favorable opinion held of him by such fellows of King's College as the economists J. M. Keynes and A. C. Pigou, despite the fact that his dissertation on the central limit theorem (a fundamental result in the theory of probability) was an act of rediscovery.

33. Dasgupta, op. cit., Chapter 3.

34. A comprehensive history of microprogramming has yet to be written. For brief accounts of the history and evolution of this topic see, e.g., S. S. Husson, *Microprogramming: Principles and Practices*, Prentice-Hall (Englewood Cliffs, N.J., 1970), pp. 22–32; S. Dasgupta, "The Organization of Microprogram Stores," *ACM Computing Surveys*, 11, no. 1 (Mar. 1979), 39–65; M. V. Wilkes, "The Growth of Interest in Microprogramming: A Literature Survey," *ACM Computing Surveys*, 1, no. 3 (1969), 139–145.

35. These conferences are the Annual Workshops on Microprogramming. For an account of the history of these workshops, see S. Habib, "A Brief Chronology of Microprogramming Activity," in S. Habib (ed.), *Microprogramming and Firmware Engineering Methods*, van Nostrand-Rheinhold (New York, 1988), pp. 1–32.

36. Husson, op. cit.

37. S. Smiles, *Industrial Biography: Iron Workers and Tool Makers*, David & Charles (Newton Abbot, U.K., 1967; 1st edition, John Murray, London, 1863).

38. L. Mumford, *Technics and Civilization*, Harcourt, Brace & World (New York, 1963, Harbinger Books edition), pp. 134, 164.

39. McGannon, op. cit., pp. 21–23.

40. Gale, op. cit., p. 159.

41. K. C. Barraclough, "Benjamin Huntsman, 1704–1776," Sheffield City Libraries,

Local Studies Leaflet, 1976, p. 1. For considerable details of the history of the crucible process, see also idem, *Steelmaking Before Bessemer, Vol. 2: Crucible Steel*, Metals Society (London, 1984), especially Chapter 1, "The Early Development of the Crucible Process," pp. 1–29.

42. This story apparently was told in *The Useful Metals and Alloys*, written anonymously and published in London in 1857. It was recounted by Smiles, op. cit., p. 108. Barraclough (1976), op. cit., p. 7 is, however, somewhat skeptical about the veracity of this incident.

43. Barraclough (1984), op. cit., p. 26.

44. H. Petroski, *Design Paradigms: Case Histories of Error and Judgement in Engineering*, Cambridge University Press (New York, 1994), Chapter 7.

45. S. P. Timoshenko, *History of Strength of Materials*, McGraw-Hill (New York, 1957; reprint, Dover [New York, 1983]), pp. 156–162.

46. N. Rosenberg and W. G. Vincenti, *The Britannia Bridge: The Generation and Diffusion of Technological Knowledge*, MIT Press (Cambridge, Mass., 1978).

47. See, e.g., Petroski, op. cit.; D. P. Billington, *The Tower and the Bridge*, Basic Books (New York, 1983), pp. 47–48; 56–59, M. Chrimes, *Civil Engineering 1839–1889*, Alan Sutton/Thomas Telford (Shroud, Gloucestershire/London, 1991); I. McNeil, "Roads, Bridges and Vehicles," in I. McNeil (ed.) *Encyclopedia of the History of Technology*, Routledge (London and New York, 1990), pp. 431–473, See, especially, p. 464.

48. The Conway Bridge, spanning the Conway River about fifteen miles from the site of the Britannia Bridge, and also built by Stephenson, was a slightly smaller version of the latter. The design and construction of the two bridges were actually overlapped in time. The principal sources of details on the building of both bridges are E. Clark, *The Britannia and Conway Tubular Bridges*, 2 volumes, Day & Sons/John Weale (London, 1850); and W. Fairbairn, *An Account of the Construction of the Britannia and Conway Tubular Bridges*, John Weale/Longman, Brown, Green and Longmans (London, 1849).

49. Billington, op. cit., p. 48; Petroski, op. cit.; Timoshenko, op. cit., pp. 160–161.

50. Rosenberg and Vincenti, op. cit., p. 60.

51. Timoshenko, op. cit., p. 160.

52. Rosenberg and Vincenti, op. cit., p. 49–50.

53. Rosenberg and Vincenti, op. cit., p. 67.

54. See, e.g., H. Petroski, *The Evolution of Useful Things*, Alfred A. Knopf (New York, 1992), Chapter 1, for a discussion of the evolution of forks.

55. E. Ferguson, *Engineering and the Mind's Eye*, MIT Press (Cambridge, Mass., 1992), p. 4; Basalla, op. cit., pp. 88–89.

Chapter 6

1. H. A. Simon, *Administrative Behavior*, 3rd edition, Free Press (New York, 1976; 1st edition 1946); idem, "Theories of Bounded Rationality" in C. B. Radner and R. Radner (eds.), *Decision and Organization*, North-Holland (Amsterdam, 1972), pp. 161–176, reprinted in H. A. Simon, *Models of Bounded Rationality*, Vol. 2, MIT Press (Cambridge, Mass., 1982), pp. 608–623; idem, *Reason in Human Affairs*, Basil Blackwell (Oxford, 1983).

2. See, e.g., Simon (1983), op. cit., pp. 12–17, for a concise summary of the model of perfect rationality; see also Simon (1972), op. cit.

3. H. Petroski, *To Engineer is Human: The Role of Failure in Successful Design*, St. Martin's Press (New York, 1985), pp. 43–46.

4. See D. C. Brown and B. Chandrasekaran, *Design Problem Solving*, Pitman (London, 1989); B. Chandrasekaran, "Design Problem Solving: A Task Analysis," *AI Magazine*, 11 (1990), 59–71.

5. S. Dasgupta, *Design Theory and Computer Science*, Cambridge University Press (Cambridge, 1991), Chapter 5.

6. S. Dasgupta, "Two Laws of Design," *Intelligent Systems Engineering*, 1, no. 2 (Winter 1992), pp. 146–156.

7. The argument leading to the derivation of the hypothesis law is presented in some detail in Dasgupta (1992), op. cit.

8. K. R. Popper, *The Logic of Scientific Discovery*, Harper and Row (New York, 1968).

9. P. B. Medawar, "Is the Scientific Paper a Fraud?" *Listener*, 70 (Sept. 12, 1963); Reprinted in P. B. Medawar, *The Threat and the Glory: Reflections on Science and Scientists*, Oxford University Press (Oxford, 1990), 228–233.

10. R. W. Cahn, "Modern Practice in the Design of Strong Alloys," *Recent Developments in Metallurgical Science and Technology, Vol. 4: Physical Metallurgy*, Indian Institute of Metals Silver Jubilee Symposium, (New Delhi, 1972), 527–551.

11. See, e.g., R. E. Reed-Hill, *Physical Metallurgy Principles*, 2nd edition, D. Van Nostrand Company (New York, 1973), especially Chapter 9, "Precipitation Hardening" for more on this phenomenon.

12. Cahn, op. cit., p. 531.

13. Cahn, op. cit., p. 531.

14. Ni_3Ti and Ni_3Al are instances of what are called "intermetallic compounds." Like all compounds, they are formed as a result of definite numbers of atoms of the constituent (metallic) elements combining with one another. For example, the formula Ni_3Ti signifies that this compound consists of three atoms of nickel combined with one atom of titanium. Similarly, Ni_3Al is an intermetallic compound consisting of three atoms of nickel and one atom of aluminum.

15. The "lattice parameter" is a spatial property of the lattice structure of crystals. "Coherency" refers to the meeting of two crystal structures (e.g., that of the matrix and of the precipitate) along a planer interface that is common to the lattices of the two structures.

16. E. Kant and A. Newell, "Problem Solving Techniques for the Design of Algorithms," *Information Processing and Management*, 20, nos. 1–2 (1986), 97–118; see also E. Kant, "Understanding and Automating Algorithm Design," *IEEE Transactions on Software Engineering*, SE-11, no. 11 (Nov. 1985), 1361–1374.

17. Such protocol analysis as a technique for studying human problem solving in a laboratory-like environment had been used extensively over a decade before by Newell and Herbert Simon in their studies of how humans play chess, solve cryptarithmetic puzzles, and undertake problems in symbolic logic. See A. Newell and H. A. Simon, *Human Problem Solving*, Prentice-Hall (Englewood Cliffs, N.J., 1972).

18. The subject S2's comments shown here are all based on the edited protocol presented by Kant and Newell, op. cit., p. 102.

19. S. Dasgupta, "Testing the Hypothesis Law of Design: The Case of the Britannia Bridge," *Research in Engineering Design*, 6, no. 1 (1994), 38–57.

20. For example, Nathan Rosenberg and Walter Vincenti have written on the role that the development of the Britannia Bridge played in the growth of engineering knowledge in the nineteenth century—see their *The Britannia Bridge: The Generation and*

Diffusion of Technological Knowledge, MIT Press (Cambridge, Mass., 1978). Stephen Timoshenko in his *History of Strength of Materials*, McGraw-Hill (New York, 1953; reprint, Dover [New York, 1983]) has also dwelt on this project in the context of how it contributed to the engineering science of strength of materials. And, most recently, Henry Petroski has discussed how the Britannia Bridge serves as an instance or as a paradigm of technological failure. See H. Petroski, *Design Paradigms: Case Histories of Error and Judgement in Engineering*, Cambridge University Press (New York, 1994), Chapter 7. See also his article "The Britannia Tubular Bridge," *American Scientist*, May–June 1992, 220–224.

21. E. Clark, *The Britannia and Conway Tubular Bridges*, 2 volumes, Day & Sons (London, 1850).

22. W. Fairbairn, *An Account of the Construction of the Britannia and Conway Tubular Bridges*, John Weale/Longman, Brown, Green and Longman (London, 1849).

23. Dasgupta, op. cit.

24. R. Stephenson, "Introductory Observations on the History of the Design," in Clark, op. cit., pp. 13–36.

25. Stephenson, op. cit., p. 25.

26. Rosenberg and Vincenti, op. cit., p. 7.

27. W. Addis, *Structural Engineering: The Nature of Theory and Design*, Ellis Horwood (Chichester, U.K., 1990), pp. 136–137.

28. Stephenson, op. cit., p. 27.

29. See, for example, M. B. Hesse, *Models and Analogies in Science*, Sheed and Ward (London, 1966), for a discussion of the role of analogy in scientific reasoning. J. Holland et al., *Induction*, MIT Press (Cambridge, Mass., 1986), especially Chapter 10, is a detailed discussion of the general nature of analogical reasoning. In S. Dasgupta, *Creativity in Invention and Design*, Cambridge University Press (New York, 1994), I discuss how inference by analogy can be used in the act of invention (pp. 161–167).

30. Stephenson, op. cit., p. 26.

31. See S. Timoshenko, *History of Strength of Materials*, Dover (New York, 1983), pp. 121–126, for a brief account of Fairbairn's work.

32. Stephenson, op. cit., p. 31.

33. See Timoshenko, op. cit., pp. 126–129 for a discussion of Hodgkinson's background and work.

34. Rosenberg and Vincenti, op. cit., p. 13.

35. Fairbairn, op. cit., p. 39. *Buckling* is a mode of structural failure in which a structural element such as a beam or a thin column that is compressed by forces on both ends exhibits sudden and catastrophic sideways bending. The reader can easily grasp the nature of this phenomenon by pushing sufficiently hard on the two ends of a common drinking straw.

36. Rosenberg and Vincenti, op. cit., pp. 19–22.

37. K. R. Popper, *The Logic of Scientific Discovery*, Harper & Row (New York, 1968). Perhaps the most well known implicit criticism of Popper's "falsificationism" is T. S. Kuhn's *The Structure of Scientific Revolutions*, 2nd edition, University of Chicago Press (Chicago, 1970; 1st edition 1962). Kuhn criticized Popper in more explicit terms in his paper "Logic of Discovery or Psychology of Research," in I. Lakatos and A. Musgrave (eds.), *Criticism and the Growth of Knowledge*, Cambridge University Press (Cambridge, 1970), pp. 1–24. For another critical view of Popper, see I. Lakatos, "Falsification and the Methodology of Scientific Research Programmes," in Lakatos and Musgrove, op. cit., pp. 91–196.

38. S. Dasgupta, "Testing the Hypothesis Law of Design: The Case of the Britan-

nia Bridge," *Research in Engineering Design*, 6, no. 1 (1994), 38–57. See especially pp. 44–47.

39. Fairbairn, op. cit., p. 18.

40. Fairbairn, op. cit., pp. 2–3.

41. Clark, op. cit., p. 478.

42. Clark, op. cit., pp. 155–157.

43. Dasgupta, op. cit., especially Fig. 12.

44. W. G. Vincenti, *What Engineers Know and How They Know It*, Johns Hopkins University Press (Baltimore, 1990), pp. 7–8.

45. Kuhn, op. cit., Chapter 2.

46. See Dasgupta, op. cit., especially pp. 48–53.

47. See Dasgupta, op. cit., p. 53 for the reasoning behind the introduction of cells in the bottom flange.

Chapter 7

1. K. R. Popper, *The Logic of Scientific Discovery*, Harper & Row (New York, 1968).

2. R. W. Weisberg, *Creativity: Genius and Other Myths*, W. H. Freeman (New York, 1986); *Creativity: Beyond the Myth of Genius*, W. H. Freeman (New York, 1993).

3. See B. Ghiselin, *The Creative Process: A Symposium*, University of California Press (Berkeley, 1952), for instances of such introspective accounts by Einstein, van Gogh, D. H. Lawrence, Wordsworth, Yeats, and other luminaries from the domains of literature, music, art, and science. R. Harding's *An Anatomy of Inspiration*, W. Heffer & Sons (Cambridge, 1942) is also a rich source of quotations, especially by writers, composers, and artists expressing wonder at their own acts of creation.

4. J. Hadamard, *The Psychology of Invention in the Mathematical Field*, Princeton University Press (Princeton, N.J., 1945; reprint Dover [New York, 1954]).

5. This lecture was, apparently, delivered in the form of a banquet speech in 1891, the banquet commemorating Helmholtz's seventieth birthday. See Graham Wallas, *The Art of Thought*, Harcourt, Brace, Jovanovich (New York, 1926).

6. H. Poincaré, "Mathematical Creation," in *The Foundations of Science*, translated by G. B. Haltstead (1908; reprint, Science Press [New York, 1952]).

7. Wallas, op. cit.

8. C. Patrick, "Creative Thoughts in Poets," in R. Woodworth (ed.) *Archives of Psychology*, 178 (1935); "Creative Thoughts in Artistic Activity," *Journal of Psychology*, 4 (1937), 35–73.

9. See, e.g., Ghiselin, op. cit.

10. R. W. Weisberg, *Creativity: Beyond the Myth of Genius*, W. H. Freeman (New York, 1993), pp. 42–50.

11. W. A. Mozart, "A letter"; reprinted in Ghiselin, op. cit., pp. 44–45.

12. J. L. Lowes, *The Road to Xanadu*, Houghton Mifflin (Boston, 1927).

13. Weisberg, op. cit., pp. 48–49.

14. Poincaré, op. cit., p. 394; italics in the original.

15. J. Gardner, *The Art of Fiction*, Vintage Books (New York, 1991), pp. 69–70.

16. L. Jeffrey, "Writing and Rewriting Poetry: William Wordsworth," in D. B. Wallace and H. E. Gruber, *Creative People at Work*, Oxford University Press (New York, 1989), pp. 69–89.

17. Jeffrey, op. cit. See, especially, her discussion of the "soul of man" section on pp. 72–74.

18. Patrick, 1937, op. cit.

19. Poincaré, op. cit., p. 386; see also, Hadamard, op. cit., pp. 29–30.

20. A. Koestler, *The Act of Creation*, Hutchinson & Co. (London, 1964; reprint, Arkana [London, 1989]).

21. Koestler, op. cit., p. 78; italics in the original.

22. S. Dasgupta, *Creativity in Invention and Design*, Cambridge University Press (New York, 1994).

23. W. B. Carlson and M. E. Gorman, "A Cognitive Framework to Understand Technological Creativity: Bell, Edison and the Telephone," in R. J. Weber and D. N. Perkins (eds.), *Inventive Minds*, Oxford University Press (New York, 1992), pp. 48–79.

24. T. D. Crouch, "Why Wilbur and Orville? Some Thoughts on the Wright Brothers and the Process of Invention," in Weber and Perkins, op. cit., pp. 80–92. The quotation is from p. 86.

25. D. T. Campbell, "Blind Variation and Selective Retention in Creative Thoughts as in Other Knowledge Processes," *Psychological Review*, 67 (1960), pp. 380–400.

26. D. T. Campbell, "Evolutionary Epistemology," in P. A. Schlip (ed.), *The Philosophy of Karl Popper*, Open Court Press (LaSalle, Ill., 1974). M. Ruse, *Taking Darwin Seriously*, Basil Blackwell (Oxford, 1986), especially Chapter 2, "Evolutionary Epistemology."

27. D. K. Simonton, *Scientific Genius: A Psychology of Science*, Cambridge University Press (Cambridge, 1988), especially, Chapter 1, "The Chance-Configuration Theory."

28. W. G. Vincenti, *What Engineers Know and How They Know It*, Johns Hopkins University Press (Baltimore, 1990), especially Chapter 8, "A Variation-Selection Model for the Growth of Engineering Knowledge."

29. Simonton, op. cit., pp. 6–16.

30. Simonton, op. cit., p. 6.

31. Simonton, ibid.

32. Simonton, op. cit., p. 7.

33. This is my example, not Simonton's.

34. B. Ghiselin, *The Creative Process*, University of California Press (Berkeley, 1952).

35. E. W. Constant II, *Origins of the Turbojet Revolution*, Johns Hopkins University Press (Baltimore, 1980), p. 10.

36. Vincenti, op. cit., p. 7.

37. Vincenti, op. cit., p. 247.

38. Vincenti, op. cit., Chapter 2.

39. Vincenti, op. cit., p. 250.

40. H. E. Gruber, *Darwin on Man: A Psychological Study of Scientific Creativity*, 2nd edition, University of Chicago Press (Chicago, 1981).

41. M. Ruse, *Taking Darwin Seriously*, Basil Blackwell (Oxford, 1986), pp. 16–17; E. Mayr, *Towards a New Philosophy of Biology*, Belknap Press of Harvard University Press (Cambridge, Mass., 1988), pp. 220–232; J. Maynard Smith, *On Evolution*, Edinburgh University Press (Edinburgh, 1972), pp. 82–91.

42. R. W. Cahn, "Modern Practice in the Design of Strong Alloys," *Recent Developments in Metallurgical Science and Technology, Vol. 4*, Indian Institute of Metals, Silver Jubilee Symposium (New Delhi, 1972), pp. 527–551.

43. Cahn, op. cit., p. 530.

44. E. Kant and A. Newell, "Problem Solving Techniques for the Design of Algorithms," *Information Processing and Management*, 20, nos. 1–2 (1984), pp. 97–118.

45. Kant and Newell, op. cit., p. 101.

46. Vincenti, op. cit., Chapter 4, pp. 112–136.

47. W. G. Vincenti, "The Retractable Landing Gear and the Northrop 'Anomaly': Variation-Selection and the Shaping of Technology," *Technology and Culture*, 35, no. 1 (Jan. 1994), 1–33.

48. R. Stephenson, "Introductory Observations on the History of the Design," Chapter 1 in E. Clark, *The Britannia and Conway Tubular Bridges*, Day & Sons/John Weale (London, 1850), p. 23.

49. Stephenson, ibid.

50. A. Newell and H. A. Simon, "Elements of a Theory of Human Problem Solving," *Psychological Review*, 65 (1958), 151–166.

51. For a comprehensive survey of the information-processing paradigm in cognitive psychology and of Newell and Simon's contribution to it, see R. Lachman, J. L. Lachman, and E. C. Butterfield, *Cognitive Psychology and Information Processing*, Lawrence Erlbaum Associates (Hillside, N.J., 1979), especially Chapter 4.

52. A. Newell, J. C. Shaw, and H. A. Simon, "The Process of Creative Thinking," in H. E. Gruber, G. Terrell, and M. Wertheimer (eds.), *Contemporary Approaches to Creative Thinking*, Atherton Press (New York, 1963), pp. 63–119.

53. Newell, Shaw, and Simon, op. cit., p. 64.

54. S. Dasgupta, *Creativity in Invention and Design*, Cambridge University Press (New York, 1994). See, especially, Chapter 2. In this work, I refer to this hypothesis as a "computational theory of scientific creativity."

55. D. Kulkarni and H. A. Simon, "The Processes of Scientific Discovery: The Strategy of Experimentation," *Cognitive Science*, 12 (1988), 139–176.

56. P. Langley, H. A. Simon, G. L. Bradshaw, and J. M. Zytkow, *Scientific Discovery*, MIT Press (Cambridge, Mass., 1987).

57. P. Thagard, *Computational Philosophy of Science*, MIT Press (Cambridge, Mass., 1988); P. Thagard, *Conceptual Revolutions*, Princeton University Press (Princeton, N.J., 1992).

58. W. B. Carlson and M. E. Gorman, "A Cognitive Framework to Understand Technological Creativity: Bell, Edison and the Telephone," in R. J. Weber and D. N. Perkins (eds.), *Inventive Minds*, Oxford University Press (New York, 1992), pp. 48–79.

59. R. J. Weber, "Stone Age Knife to Swiss Army Knife: An Invention Prototype," in Weber and Perkins, op. cit., pp. 217–237.

60. Dasgupta, op. cit., especially Chapters 5 and 6.

61. For a discussion of the invention of microprogramming as a historically original act, see Dasgupta, op. cit., pp. 24–26.

62. Details of these articles, lectures, and "other sources" are given in Dasgupta, op. cit.

63. See, in particular, Dasgupta, op. cit., Chapters 5 and 6.

64. A stored-program computer is a computer in which the computer program is automatically loaded into the computer memory and then executed by the computer without manual intervention. Prior to the Cambridge and Manchester machines, a program had to be manually introduced into the computer by setting switches and plugging in cables. This was the case, for example, with the celebrated ENIAC completed in 1946 at the University of Pennsylvania.

65. R. J. Sternberg, "A Three-Facet Model of Creativity," in R. J. Sternberg (ed.),

The Nature of Creativity, Cambridge University Press (Cambridge, 1988), pp. 125–147. See, especially, pp. 132–133.

66. Sternberg, op. cit., pp. 133–134.

67. M. V. Wilkes, *Memoirs of a Computer Pioneer*, MIT Press (Cambridge, Mass., 1985), p. 128; "The Genesis of Microprogramming," *Annals of the History of Computing*, 8, no. 2 (1986), 117; see Dasgupta, op. cit., pp. 75–77 for a description of how the diode matrix is used in the EDSAC.

68. Wilkes (1985), op. cit., Chapter 16; Wilkes (1986), op. cit.

69. The labels or identities for facts, rules, and goals shown here, e.g., *F4*, *R2*, *G2*, etc., are those used in the original description of this process—see Dasgupta, op. cit.

70. Wilkes (1985), op. cit., p. 178, Wilkes (1986), op. cit., p. 117.

71. *Abduction* is a kind of reasoning, originally identified by the nineteenth-century philosopher C. S. Peirce, that allows one to infer a cause from the observation of an effect. Suppose some phenomenon or problem *P* is observed, and it is known that *P* would be explained or solved if a situation *S* happens to be the case. Then by abductive reasoning, it may be inferred that *S* is the case.

72. J. S. Mill, *A System of Logic: Ratiocinative and Inductive* (1843; reprint, University of Toronto Press [Toronto, 1974]).

73. J. H. Holland, K. J. Holyoak, R. E. Nisbett, and P. R. Thagard, *Induction*, MIT Press (Cambridge, Mass., 1986).

74. P. R. Thagard, *Computational Philosophy of Science*, MIT Press (Cambridge, Mass., 1988).

75. R. Stephenson, "Introductory Observations on the History of the Design," Chapter 1 in E. Clark, *The Britannia and Conway Tubular Bridges*, Day & Sons/John Weale (London, 1850), pp. 21 et seq.

76. Stephenson, ibid.

77. This same rule was in fact employed to initiate the postulated process leading to Wilkes's invention of microprogramming. See Dasgupta, op. cit., p. 104.

78. *Modus ponens* is an inference rule in deductive logic which states that if the premises of "*A*" and "*A* implies *B*" hold, then one may conclude that "*B*" holds.

79. Stephenson, op. cit.

80. Stephenson, ibid.

81. Stephenson, ibid.

82. Stephenson, ibid.

83. Stephenson, ibid.

84. A. Newell and H. A. Simon, *Human Problem Solving*, Prentice-Hall (Englewood Cliffs, N.J., 1972).

85. In S. Dasgupta, *Creativity in Invention and Design* (op. cit., p. 168), this form of the means-ends heuristic is called the "misfit identification and elimination rule."

86. Stephenson, op. cit., p. 24.

87. A. Newell, *Unified Theories of Cognition*, Harvard University Press (Cambridge, Mass., 1990), pp. 98–99.

88. Newell and Simon (1972), op. cit., pp. 59–78.

89. Newell (1990), op. cit.

90. See, e.g., Dasgupta, op. cit., pp. 27–33, for a discussion of the role of metaphor in creative problem solving.

91. See, e.g., S. Dasgupta, *Design Theory and Computer Science*, Cambridge University Press (Cambridge, 1991), especially Chapter 9, where a method called 'plausibility-driven design' is described. In this method, the designer explicitly records the design rationale as an integral part of the design process.

92. *Bandwidth* is one of the established performance measures for computer systems. Broadly speaking, it is defined as the amount of information transmitted across a communication channel connecting two key components. For instance, *memory bandwidth* refers to the amount of information transmitted per unit time between memory and the processing unit in a computer. *I/O bandwidth* refers to the amount of information transmitted per unit time by an I/O system to a computer. *Latency* refers to the time delay from when a certain task is initiated to when it is completed. Thus, *I/O latency* is a measure of the delay between the initiation of an I/O operation and its completion. *Throughput*, another performance measure, is the number of tasks or operations that can be processed by a system per unit time. A common measure of throughput, in the case of a computer, is the number of millions of instructions executed by the computer per second.

93. Newell, op. cit., p. 102.

94. Newell, op. cit., p. 99.

95. K. C. Barraclough, *Steelmaking before Bessemer, Vol. 2, Crucible Steel*, Metals Society (London, 1986), p. 1; K. C. Barraclough, "Benjamin Huntsman, 1704–1776; Sheffield City Libraries, Local Studies Booklet, 1976.

96. It is interesting to note that John Gardner, in *The Art of Fiction* (Vintage Books, New York, 1991) makes the point (pp. 10–11) that the creative writer must, in a literary sense, be knowledgeable. "No ignoramus," he writes, "has ever produced great art" (p. 10). For, among other things, "All great writing is in a sense imitation of great writing" (p. 11).

Chapter 8

1. S. J. Gould, *Ontogeny and Phylogeny*, Belknap Press of Harvard University Press (Cambridge, Mass., 1977), p. 483.

2. See S. Dasgupta, *Design Theory and Computer Science*, Cambridge University Press (Cambridge, 1991), p. 114.

3. Dasgupta, op. cit., p. 115.

4. Gould, op. cit., p. 483.

5. J. P. Steadman, *The Evolution of Designs*, Cambridge University Press (Cambridge, 1979).

6. Steadman, op. cit., pp. 87–102.

7. G. Basalla, *The Evolution of Technology*, Cambridge University Press (Cambridge, 1988).

8. H. Petroski, *The Evolution of Useful Things*, Alfred A. Knopf (New York, 1992).

9. E. J. Hobsbawm, *The Age of Revolution*, Weidenfeld and Nicolson (London, 1962; reprint, Cardinal [London, 1973]), p. 44. I. B. Cohen, *Revolution in Science*, Belknap Press of Harvard University Press (Cambridge, Mass., 1985), pp. 265–266.

10. D. S. L. Cardwell, *Turning Points in Western Technology*, Science History Publications (New York, 1972).

11. Basalla, op. cit., p. 35; H. W. Dickinson, *A Short History of the Steam Engine*, Cambridge University Press (Cambridge, 1939), pp. 68–69; A. P. Usher, *A History of Mechanical Inventions*, revised edition, Harvard University Press (Cambridge, Mass., 1954; reprint Dover [New York, 1988]; 1st edition 1929), p. 353.

12. In his monograph on Newcomen and the early history of the steam engine, the historian of technology L. T. C. Rolt has pointed out that in Newcomen's time an ironmonger was not merely a supplier of ironware but also a maker of the goods he

sold. Perhaps a more accurate description of Newcomen's occupation would be blacksmith or even foundryman. L. T. C. Rolt, *Thomas Newcomen*, David and Charles/ Dawlish MacDonald (London, 1963), p. 48.

13. Rolt, op. cit., p. 10; Usher, op. cit., p. 347; R. H. Thurston, *A History of the Growth of the Steam Engine*, Cornell University Press (Ithaca, N.Y., 1939; 1st edition 1878), pp. 56–57. It must be kept in mind, however, that in Newcomen's engine, power was derived primarily from the force of the atmosphere rather than the force of steam. Thus, strictly speaking, Newcomen's "atmospheric" engine constituted the high point of the *prehistory* of the steam engine.

14. Thurston, op. cit., p. 58; Usher, op. cit., p. 348; Rolt, op. cit., p. 52.

15. Dickinson, op. cit., p. 32.

16. Cardwell, op. cit., p. 55; Dickinson, op. cit., p. 7.

17. Cardwell, op. cit., p. 56.

18. Dickinson, op. cit., p. 9.

19. Dickinson, op. cit., p. 10.

20. Cardwell, op. cit., p. 56; Dickinson, op. cit., p. 12.

21. Dickinson, op. cit., p. 12–13; Rolt, op. cit., p. 21; Thurston, op. cit., p. 15.

22. Dickinson, op. cit., p. 13.

23. Quoted by Dickinson, op. cit., p. 13.

24. Dickinson, op. cit., p. 13.

25. Usher, op. cit., pp. 343–344.

26. Thurston, op. cit., pp. 21–22; Usher, op. cit., pp. 343–345.

27. Rolt, op. cit., p. 26.

28. Rolt, op. cit., p. 27.

29. Rolt, op. cit., p. 27.

30. Thurston, op. cit., pp. 28–29.

31. Cardwell, op. cit., p. 59.

32. Usher, op. cit., p. 346.

33. Quoted by Dickinson, op. cit., p. 20.

34. *Philosophical Transactions of the Royal Society*, 21, no. 253 (1689), p. 228. See also, Thurston, op. cit., p. 33; Dickinson, op. cit., pp. 22–23.

35. The description and diagram presented here are based on the account given by Thurston, op. cit., pp. 34–35.

36. Thurston, op. cit., pp. 57–58.

37. Rolt, op. cit., pp. 49–50.

38. Thurston, op. cit., p. 57–58; Dickinson, op. cit., p. 32.

39. Thurston, op. cit., p. 58; Usher, op. cit., p. 348.

40. Thurston, op. cit., p. 58.

41. Usher, op. cit., p. 348.

42. Dickinson, op. cit., pp. 32–33, Rolt, op. cit., p. 49.

43. Usher, op. cit., p. 347. There is, however, an interesting ideological tension between such earlier statements as that by Robison and the recent accounts by Rolt and Dickinson that is worth noting. Early in his book, Rolt makes the point that one of the reasons why Newcomen's achievement and fame was undermined by the luminaries of the day was the huge intellectual gap that was perceived to exist between the "natural philosophers"—the pure scientists—and the practical makers—the smiths, ironworkers, and so on. It was inconceivable to the scientists and other intellectuals of the time, specifically the eighteenth century, that someone of Newcomen's relatively humble background would conceive such an artifact as the atmospheric engine. They attributed his success to the work that had come before him or to chance. [See Rolt,

op. cit., pp. 11–12, 49–50; Dickinson, op. cit., p. 36.] Rolt himself seemed anxious to counter this bias. That is, his history of the Newcomen engine appears to deemphasize, very deliberately, Newcomen's debt to the past, in particular, to the possibility of his having been aware of Papin's and Savory's inventions.

44. Dickinson, op. cit., pp. 32–33.

45. For further discussion of the general purpose nature of the ENIAC in contrast to the more specialized character of its antecedents, see A. W. Burks and A. R. Burks, "The ENIAC: First General Purpose Electronic Computer," *Annals of the History of Computing*, 3, no. 4 (Oct. 1981), 310–399, especially p. 385.

46. See Burks and Burks, op. cit., pp. 312–313, 389–392. The records of the trial are available as *The ENIAC Trial Records*, U.S. District Court, District of Minnesota, Fourth Division: Honeywell Inc. vs. Sperry Rand Corp. *et al*, No. 4–67, Civ. 138, decided Oct. 19, 1973: Judge Earl Larson. Judge Larson's decision was printed in *U.S. Patent Quarterly*, 180 (March 25, 1974), 673–773.

47. I will be relying, in particular, on the following sources: For historical and retrospective personal accounts, Burks and Burks, op. cit.; A. R. Burks, "From ENIAC to the Stored Program Computer: Two Revolutions in Computers," in N. Metropolis, J. Howlett, and G.-C. Rota (eds.), *A History of Computing in the Twentieth Century*, Academic Press (New York, 1980), pp. 311–344; J. P. Eckert, "The ENIAC," in Metropolis, Howlett, and Rota, op. cit., pp. 525–540; J. W. Mauchly, "The ENIAC" in Metropolis, Howlett, and Rota, op. cit., pp. 541–550; C. R. Mollenhoff, *Atanasoff: Forgotten Father of the Computer*, Iowa State University Press (Ames, 1988); H. H. Goldstine, *The Computer from Pascal to von Neumann*, Princeton University Press (Princeton, N.J., 1972); J. V. Atanasoff, "Advent of Electronic Digital Computing," *Annals of the History of Computing*, 6, no. 3 (Jul. 1984), 229–282; J. R. Berry, "Clifford E. Berry, 1918–1963: His Role in Early Computers," *Annals of the History of Computing*, 8, no. 4 (Oct. 1986), 361–369; K. K. Mauchly, "John Mauchly's Early Years," *Annals of the History of Computing*, 6, no. 2 (Apr. 1984), 116–138.

For original papers on the relevant projects: J. V. Atanasoff, "Computing Machine for the Solution of Large Systems of Linear Algebraic Equations," unpublished memorandum, Iowa State College, Ames, Aug. 1940, reprinted in B. Randell (ed.), *The Origins of Digital Computers, Selected Papers*, 2nd edition, Springer-Verlag (New York, 1975), pp. 305–326, J. W. Mauchly, "The Use of High Speed Vacuum Tubes for Calculating," unpublished memorandum, University of Pennsylvania, Philadelphia, Aug. 1942, reprinted in Randell, op. cit., pp. 329–332; H. H. Goldstine and A. Goldstine, "The Electronic Numerical Integrator and Computer (ENIAC)," *Mathematical Tables and Automatic Computation*, 2, no. 15 (1946), 97–110, reprinted in Randell, op. cit., pp. 333–347.

48. In the terminology of computers, an *accumulator* is a storage device capable of holding or storing a single number of some fixed size (e.g., a signed ten digit number as in the case of the ENIAC) which has a built-in adder that allows the addition (or subtraction) of some input numbers to (or from) the contents of the accumulator and storing the result back into the accumulator.

49. This example is taken from Goldstine, op. cit., p. 160.

50. Burks and Burks, op. cit.

51. Burks and Burks, op. cit., p. 328; Atanasoff, op. cit., pp. 239, 241–247; Berry, op. cit., p. 366.

52. Atanasoff, op. cit., p. 255; Burks and Burks, op. cit., p. 331.

53. V. Bush, "The Differential Analyzer, a New Machine for Solving Differential Equations," *Journal of the Franklin Institute*, 212 (1931), 447–488.

54. Burks and Burks, op. cit., p. 315.

55. Larson, op. cit.; Burks and Burks, op. cit., p. 312.

56. Atanasoff's original description of his machine was the unpublished 1940 memorandum, Atanasoff (1940), op. cit. For a more recent account by him, see Atanasoff (1984), op. cit.

57. Burks and Burks, op. cit., p. 317; Atanasoff (1984), op. cit., p. 242.

58. In a *binary* computer, all information—numbers or otherwise—is encoded and stored as patterns of binary digits, that is, as patterns of the digits 0 and 1. All computational circuits, such as those responsible for addition, subtraction, or logical operations, are designed to manipulate such binary strings. In the ABC, a decimal number such as 21, punched on a card, would be converted, according to a conversion code, into a corresponding and unique binary form (10101) before being stored inside the machine. Conversely, if the result of a computation is, say, 111000, this would be converted into its decimal form (56) before being punched onto an output card.

59. Burks and Burks, op. cit., p. 330.

60. Atanasoff (1984), op. cit., p. 255, Burks and Burks, op. cit., p. 331.

61. Mollenhoff, op. cit., p. 94.

62. Mollenhoff, op. cit., pp. 55–58, Atanasoff (1984), op. cit., pp. 254–255.

63. Mollenhoff, op. cit., p. 57. The report referred to here is Atanasoff (1940), op. cit.

64. Mollenhoff, op. cit., p. 59.

65. Mauchly (1940), op. cit.

66. Burks, op. cit., p. 314. See also Goldstine, op. cit., pp. 148–149.

67. Mauchly (1942), op. cit.; reprinted in Randell, op. cit., p. 329.

68. See B. Randell's introduction to the chapter "Bell Telephone Laboratories," in Randell, op. cit., pp. 237–240.

69. Randell, op. cit., p. 238.

70. Burks and Burks, op. cit., p. 364.

71. This discussion is based on a detailed description of the ENIAC's high-speed multiplication unit in Burks and Burks, op. cit., pp. 362–370.

72. Burks and Burks, op. cit., p. 363.

73. Burks and Burks, op. cit., p. 369.

74. Burks and Burks, op. cit., p. 369.

75. Burks, op. cit., p. 315.

76. Goldstine, op. cit., p. 155.

77. Burks and Burks, op. cit., pp. 371–372.

78. Mollenhoff, op. cit., p. 59.

79. Mauchly (1942), op. cit.; reprinted in Randell, op. cit., pp. 329–332.

80. Burks and Burks, op. cit., p. 334.

81. Burks and Burks, op. cit., pp. 334–335.

82. Burks and Burks, op. cit., p. 344.

83. S. Dasgupta, *Design Theory and Computer Science*, Cambridge University Press (Cambridge, U.K., 1991), pp. 114–115.

84. One of the celebrated landmarks in the history of biological thought is the notion, which actually goes back to Aristotle but which was given modern shape by some nineteenth-century German biologists, notably Ernst Haeckel, that there is a parallel between ontogeny and phylogeny: that in the course of its development from embryo to adult stages, an organism passes through stages that are comparable to adult forms of the evolutionary ancestors of the organization. That, in fact, *ontogeny recapitulates phylogeny*. More technically, this is known as the "law of recapitulation." For more on this, see, e.g., P. B. Medawar and J. S. Medawar, *Aristotle to Zoos: A*

Philosophical Dictionary of Biology, Harvard University Press (Cambridge, Mass., 1983), pp. 225–226. A modern authoritative text on this law is S. J. Gould, *Ontogeny and Phylogeny*, Belknap Press of Harvard University Press (Cambridge, Mass., 1977).

85. See, e.g., S. Timoshenko, *History of Strength of Materials*, McGraw-Hill (New York, 1953; reprint, Dover [New York, 1983]), pp. 72–73, for brief remarks on the history of the suspension bridge; and pp. 181 et seq. for a longer history of trusses.

86. G. Basalla, *The Evolution of Technology*, Cambridge University Press (Cambridge, 1988).

87. E. W. Constant II, *The Origins of the Turbojet Revolution*, Johns Hopkins University Press (Baltimore, 1980).

88. T. S. Kuhn, *The Structure of Scientific Revolutions*, 2nd edition, University of Chicago Press (Chicago, 1970; 1st edition 1962).

89. Constant, op. cit., p. 241.

90. H. Petroski, *The Evolution of Useful Things*, Alfred A. Knopf (New York, 1992).

91. H. Petroski, *The Pencil: A History of Design and Circumstance*, Alfred A. Knopf (New York, 1989).

92. See Brian Randell's Introduction, pp. 1–6, and the series of excerpts of papers by Babbage and others in Chapter 2, "Analytical Engines," pp. 17–124, in B. Randell, *The Origins of Digital Computers, Selected Papers*, 2nd edition, Springer-Verlag (New York, 1975).

93. See Randell, "Analytical Engines," Chapter 2 in Randell, op. cit., p. 7–15.

94. M. V. Wilkes, "Babbage as a Computer Pioneer," *Historia Mathematica*, 4 (1977), 415–440.

95. Specifically, the Automatic Sequence Controlled Calculator-I, also called the Harvard MARK I, invented by Howard Aiken at Harvard University in collaboration with the IBM engineers C. D. Lake, F. E. Hamilton, and B. M. Durfee. This machine became operational in May 1944. See H. H. Aiken, "Proposed Automatic Calculating Machine," an unpublished memorandum written in 1937 and reprinted in Randell, op. cit., pp. 191–197; and H. H. Aiken and G. M. Hopper, "The Automatic Sequence Controlled Calculator, *Electrical Engineering*, 65 (1946), 384–391, 449–454, 522–528; reprinted in Randell, op. cit., pp. 199–218.

96. In his biography of John Atanasoff, Clark Mollenhoff quotes Presper Eckert (from a conversation held in April 1987) as saying that neither he (Eckert) nor Mauchly had ever heard of Charles Babbage. Mollenhoff, op. cit., p. 228.

97. In his retrospective account in 1984, of the circumstances leading to the invention of the ABC, Atanasoff mentions that he conducted a survey in 1935–36 of the state of the art of both analog and digital computing machines. Among the latter are mentioned desk calculators and the IBM punched-card tabulating machine. But nothing is said about Babbage. Atanasoff does mention, however, in the same paper, that he had had some knowledge of Babbage's work and that his ideas about the advantages of having a large, fast memory "were in accord with Babbage's thinking although I did not appreciate it at the time." See Atanasoff (1984), op. cit., p. 238.

Chapter 9

1. See, e.g., J. Shrager and P. Langley (eds.), *Computational Models of Scientific Discovery and Theory Formation*, Morgan Kaufmann (San Mateo, Calif., 1990).

2. T. S. Kuhn, *The Structure of Scientific Revolutions*, 2nd edition, University of

Chicago Press (Chicago, 1970; 1st edition 1962). See also L. Laudan, *Progress and its Problems*, University of California Press (Berkeley, 1977) in which Kuhn's concept of the paradigm is enlarged by Laudan to the notion of the "research tradition."

3. M. Shaw, "Prospects for an Engineering Discipline of Software," *IEEE Software*, Nov. 1990, 15–24.

4. Quoted on p. 690 by E. T. Layton, Jr., "American Ideologies of Science and Engineering," *Technology and Culture*, 17, no. 4 (Oct. 1976), 688–701.

5. Quoted by Layton, op. cit., p. 690.

6. Cited by Layton, op. cit., p. 689.

7. E. T. Layton, Jr., "Technology as Knowledge," *Technology and Culture*, 15, no. 1 (Jan. 1974), 31–41.

8. A. S. Darling, "Non-Ferrous Metals," in I. McNeil (ed.), *An Encyclopedia of the History of Technology*, Routledge (London, 1990), pp. 47–145. See especially pp. 56–58.

9. W. G. Vincenti, *What Engineers Know and How They Know It*, Johns Hopkins University Press (Baltimore, 1990), p. 227.

10. W. G. Vincenti, "Engineering Knowledge, Type of Design, and Level of Hierarchy: Further Thoughts about *What Engineers Know* . . . ," in P. Kroes and M. Bakker (eds.), *Technological Development and Science in the Industrial Age*, Kluwer Academic Publishers (Boston, 1992), pp. 17–34.

11. Walter Vincenti, in attempting to distinguish between scientific and engineering knowledge, was acutely aware of this fact. He refers to the "interpretation" of the two kinds of knowledge-generating activities. See Vincenti (1990), op. cit., p. 226.

12. C. S. Smith, "Matter versus Material: A Historical View," *Science*, 162 (1968), 637–644. Reprinted in C. S. Smith, *A Search for Structure*, MIT Press (Cambridge, Mass., 1982), 112–126.

13. Smith, op. cit., p. 113.

14. Layton (1976), op. cit., p. 695.

15. E. T. Layton, Jr., "Mirror-Image Twins: The Communities of Science and Technology in 19th Century America," *Technology and Culture*, 12, no. 4 (Oct. 1971), 562–580.

16. W. Addis, *Structural Engineering: The Nature of Theory and Design*, Ellis Horwood (Chichester, U.K., 1990), pp. 136–137.

17. M. Polanyi, *Personal Knowledge*, University of Chicago Press (Chicago, 1962), p. 176.

18. W. Fairbairn, *An Account of the Construction of the Britannia and Conway Bridges*, John Weale/Longman, Brown, Green and Longmans (London, 1849), p. 1.

19. Fairbairn, op. cit., p. 1.

20. R. Stephenson, "Introductory Observations on the History of the Design," in E. Clark, *The Britannia and Conway Tubular Bridges, Vol. 1*, Day & Son/John Weale (London, 1850), p. 21.

21. N. Rosenberg and W. G. Vincenti, *The Britannia Bridge: The Generation and Diffusion of Technological Knowledge*, MIT Press (Cambridge, Mass., 1978), pp. 19–25.

22. Vincenti, 1990, op. cit., pp. 112–136.

23. Galileo, *Dialogue Concerning Two New Sciences* (1638), English translation by H. Crew and A. deSalvio, Macmillan and Co. (New York, 1933).

24. Rosenberg and Vincenti, op. cit., p. 23.

25. Rosenberg and Vincenti, op. cit., pp. 24–25.

26. L. T. C. Rolt, *Thomas Newcomen*, David and Charles/Dawlish MacDonald (London, 1963), pp. 11–12.

27. See H. W. Dickinson, *A Short History of the Steam Engine*, Cambridge University Press (Cambridge, 1939), p. 10.

28. Dickinson, op. cit., pp. 4–5.

29. C. S. Smith, "The Discovery of Carbon in Steel," *Technology and Culture*, 5 (1965), 149–175. Reprinted in C. S. Smith, *A Search for Structure: Selected Essays on Science, Art and History*, MIT Press (Cambridge, Mass., 1981), pp. 33–53.

30. Dickinson, op. cit., p. 11.

31. C. R. Mollenhoff, *Atanasoff: Forgotten Father of the Computer*, Iowa State University Press (Ames, 1988), p. 34.

32. J. V. Atanasoff, "Computing Machine for the Solution of Large Systems of Linear Algebraic Equations," unpublished memorandum, Iowa State College (Ames, 1940). Reprinted in B. Randell (ed.), *The Origins of Digital Computers: Selected Papers*, 2nd edition, Springer-Verlag (New York, 1975), 305–326.

33. C. E. Shannon, "A Symbolic Analysis of Relay and Switching Circuits," *AIEE Transactions*, 57 (1938), 713–723. It is interesting to note, though, that Samuel H. Caldwell, who was on Shannon's M.S. thesis committee, had corresponded with Atanasoff in 1939–40, and had actually visited Iowa in the fall of 1939 to examine and review Atanasoff's project on behalf of the National Defense Research Corporation. See Mollenhoff, op. cit., pp. 119, 163.

34. Mollenhoff, op. cit., p. 34.

35. Atanasoff, op. cit., p. 309.

36. As laid out in S. Dasgupta, *Creativity in Invention and Design*, Cambridge University Press (New York, 1994), especially Chapters 5 and 6.

37. G. Sturt, *The Wheelwright's Shop*, Cambridge University Press (Cambridge, 1923, Canto edition, 1993).

38. F. P. Brooks, Jr., *The Mythical Man-Month: Essays in Software Engineering*, Addison-Wesley (Reading, Mass., 1975).

39. P. Freeman, *Software Perspectives*, Addison-Wesley (Reading, Mass., 1987).

40. M. Shaw, "Prospects for an Engineering Discipline of Software," *IEEE Software* (Nov. 1980), 15–24.

41. D. E. Knuth and R. W. Floyd, "Notes on Avoiding 'Go To' Statements," *Information Processing Letters*, 1 (1971), 23–31.

42. D. E. Knuth, "The Errors of TEX," *Software—Practice and Experience*, 19 (July 1989), 607–685. Reprinted in D. E. Knuth, *Literate Programming*, Center for the Study of Language and Information (Stanford, Calif., 1982).

43. A. Newell and H. A. Simon, *Human Problem Solving*, Prentice-Hall (Englewood Cliffs, N.J., 1972).

44. M. Polanyi, *Personal Knowledge*, University of Chicago Press (Chicago, 1962).

45. D. Schön, *The Reflective Practitioner*, Basic Books (New York, 1983).

46. Schön, op. cit., p. 21.

47. Schön, op. cit., p. 32.

48. Schön, op. cit., p. 34.

49. H. A. Simon, "The Structure of Ill-Structured Problems," *Artificial Intelligence*, 4 (1973), 181–200.

50. For extensive discussions of this aspect of the design process, see S. Dasgupta, *Design Theory and Computer Science*, Cambridge University Press (Cambridge, 1991), especially Chapters 3, 7, and 9.

51. Dasgupta, op. cit., Chapter 9.

52. Schön, op. cit., pp. 49–69; also Chapter 5.

53. The notion of tacit knowledge forms the underlying theme of Michael Polanyi's *Personal Knowledge*, op. cit. It is further discussed by Polanyi in a series of essays that

are reprinted in Part 3 of his *Knowing and Being*, University of Chicago Press (Chicago, 1969).

54. Schön, op. cit., p. 55.

55. E. Clark, *The Britannia and Conway Tubular Bridges*, Vol. 1, Day & Son/John Weale (London, 1850), p. 137.

56. N. Rosenberg and W. G. Vincenti, *The Britannia Bridge: The Generation and Diffusion of Technological Knowledge*, MIT Press (Cambridge, Mass., 1978), pp. 19–25.

57. G. Sturt, *The Wheelwright's Shop*, Cambridge University Press (Cambridge, 1923; Canto edition, 1993), p. 92.

58. Sturt, op. cit., p. 92.

59. Sturt, op. cit., p. 93.

60. Sturt, ibid.

61. Sturt, op. cit., p. 94.

62. H. Petroski, "Paradigms for Human Error in Design," *Proceedings of the 1991 NSF Design and Manufacturing Systems Conference* (Austin, Tex., 1991).

63. H. Petroski, "Paconius and the Pedestal for Apollo: A Case Study of Error in Conceptual Design," *Research in Engineering Design*, 3 (1991), 123–128. Petroski's discussion is based on Vitruvius's account of the episode in his *Ten Books on Architecture*. See also H. Petroski, *Design Paradigms: Case Histories of Error and Judgement in Engineering*, Cambridge University Press (New York, 1994), Chapter 2.

64. H. Petroski, "Galileo's Confirmation of the False Hypothesis: A Paradigm for Logical Error in Design," *Civil Engineering Systems*, 9, 1992, 251–263. See also Petroski (1994), op. cit., Chapter 5.

65. H. Petroski, "Vitruvius's Auger and Galileo's Bones: Paradigms of Limits to Size in Design," *J. Mech. Design*, 114 (Mar. 1992), 23–28. See also Petroski (1994), op. cit., Chapter 3.

66. Petroski (1994), op. cit., Chapter 7.

67. D. E. Knuth, "The Errors of TEX," *Software—Practice and Experience*, 19 (Jul. 1989), 607–685. Reprinted in D. K. Knuth, *Literate Programming*, Center for the Study of Language and Information (Stanford, Calif., 1992).

68. For an extensive discussion of the psychology of errors, see J. Reason, *Human Error*, Cambridge University Press (Cambridge, 1990).

69. B. Cotterall and J. Kamminga, *Mechanics of Pre-Industrial Technology*, Cambridge University Press (Cambridge, 1990).

70. S. P. Timoshenko, *History of Strength of Materials*, McGraw-Hill (New York, 1953; reprint, Dover [New York, 1983]), p. 62.

71. Timoshenko, op. cit., p. 63.

72. Timoshenko, op. cit., pp. 63–64.

73. Timoshenko, op. cit., p. 64.

74. H. N. Das-Gupta, "The Development, Progress and Scientific Aspects of Powder Metallurgy," Presidential Address, Section of Engineering & Metallurgy, 48th Indian Science Congress, Roorkee, India, 1961, 1–20.

75. Das-Gupta, op. cit., p. 2.

76. Das-Gupta, op. cit., p. 11.

77. Das-Gupta, op. cit., p. 12; R. W. Cahn, "Recovery and Recrystallization," in R. W. Cahn (ed.), *Physical Metallurgy*, North-Holland/ Elsevier (Amsterdam and New York, 1970), 1129–1197, especially p. 1193.

78. Rosenberg and Vincenti, op. cit., p. 21.

79. In cognitive science, the term *mental model* refers to the construction, by an agent, of appropriate internal symbolic structures that represent more or less directly those aspects of the external world that the agent is engaged with at any particular

time. To take an example due to Philip Johnson-Laird, on hearing an assertion "the dishwasher is on the right of the cupboard," the agent constructs an internal model in which symbols representing the entities "dishwasher" and "cupboard" are placed in a representation of the space such that the overall situation matches the assertion. The model has the same structure as the actual scene in the external world that prompted the assertion to be uttered. The concept of a mental model was first advanced by Kenneth Craik in *The Nature of Explanation*, Cambridge University Press (Cambridge, 1942). For a systematic discussion of the subject, see P. N. Johnson-Laird, "Mental Models," in M. I. Posner (ed.), *Foundations of Cognitive Science*, MIT Press (Cambridge, Mass., 1989), 469–499. The general role of mental models in design is discussed in D. A. Norman, *The Design of Everyday Things*, Doubleday (New York, 1989), pp. 189–190. The use of mental models in the design of a specific class of artifacts, computer systems, is discussed in S. Dasgupta, *Design Theory and Computer Science*, Cambridge University Press (Cambridge, 1991), pp. 24–27, 170–180. Both Norman and Dasgupta use the alternative term *conceptual model* as a synonym for *mental model*.

80. E. Ferguson, *Engineering and the Mind's Eye*, MIT Press (Cambridge, Mass., 1992).

81. Ferguson quotes, among others, Oliver Evans, inventor of the automatic flour mill in the 1780s; the steam engine pioneer James Watt, from the same period; the great nineteenth-century engineer Isambard Brunel; James Nasmyth, another Victorian engineer and inventor; and Walter Chrysler, the founder of the automobile company early in this century, all of whom have recorded their dependence on visual, pictorial thinking. Ferguson, op. cit., pp. 47–56.

82. Ferguson, op. cit., p. 49.

83. J. C. Jones, *Design Methods: Seeds of Human Future*, John Wiley & Sons (New York, 1981), pp. 16–24.

84. Ferguson, op. cit., pp. 90–92.

85. D. P. Billington, *Robert Maillart's Bridges: The Art of Engineering*, Princeton University Press (Princeton, N.J., 1979).

86. Billington, op. cit., pp. 91–92.

87. D. Pye, *The Nature and Aesthetics of Design*, Herbert Press (London, 1978), p. 41.

Chapter 10

1. See Chapter 4.
2. See Chapter 9.
3. See Chapter 4.
4. Ibid.
5. Ibid.
6. See Chapter 5.
7. See Chapter 7.
8. Ibid.
9. See S. Dasgupta, *Creativity in Invention and Design*, Cambridge University Press (New York, 1994), pp. 189–210, for a more detailed discussion of these characteristics in the context of the specific invention of microprogramming by Maurice Wilkes.
10. See Chapter 7.
11. See Chapter 6.
12. See Chapter 8.

Bibliography

Addis, W., *Structural Engineering: The Nature of Theory and Design*, Ellis Horwood (Chichester, U.K., 1990).

Agricola, G., *De Re Metallica* (1556), H. C. Hoover & L. H. Hoover (Tr.), Dover (New York, 1950).

Aiken, H. H., "Proposed Automatic Calculating Machine," unpublished memorandum, 1937. Reprinted in B. Randell (ed.), *The Origins of Digital Computers, Selected Papers*, 2nd edition, Springer-Verlag (New York, 1975), pp. 191–197.

Aiken, H. H., and Hopper, G. M., "The Automatic Sequence Controlled Calculator," *Electrical Engineering*, 65 (1946), 384–391, 449–454, 522–528. Reprinted in B. Randell (ed.), *The Origins of Digital Computers, Selected Papers*, 2nd edition, Springer-Verlag (New York, 1975), pp. 199–218.

Alexander, C., *Notes on the Synthesis of Form*, Harvard University Press (Cambridge, Mass., 1964).

Alexander, C., Ishikawa, S., and Silverstein, M., *A Pattern Language*, Oxford University Press (New York, 1977).

Alexander, C., Neis, H., Anninou, A., and King, I., *A New Theory of Urban Design*, Oxford University Press (New York, 1987).

Anderson, J. R., *The Architecture of Cognition*, Harvard University Press (Cambridge, Mass., 1983).

Aspray, W., *John von Neumann and the Origins of Modern Computing*, MIT Press (Cambridge, Mass., 1990).

Arbib, M. A., and Hesse, M. B., *The Construction of Reality*, Cambridge University Press (Cambridge, 1986).

Atanasoff, J. V., "Computing Machine for the Solution of Linear Algebraic Equations," unpublished memorandum, Iowa State College, Ames, Aug. 1940. Reprinted in B. Randell (ed.), *The Origins of Digital Computing, Selected Papers*, 2nd edition, Springer-Verlag (New York, 1975), pp. 305–326.

Atanasoff, J. V., "Advent of Electronic Digital Computing," *Annals of the History of Computing*, 6, no. 3 (Jul. 1984), pp. 229–282.

Atkinson, R. C., and Shiffrin, R. M., "Human Memory: A Proposed System and its Control Processes," in K. Spence and J. Spence (eds.), *The Psychology of Learning and Motivation*, Vol. 2, Academic Press (New York, 1968).

Babbage, C., *The Works of Charles Babbage*, 9 volumes, M. Campbell-Kelly (ed.), William Pickering (London, 1989).

Babbage, C., *Passages from the Life of a Philosopher*, M. Campbell-Kelly (ed.), Pickering & Chatto (London, 1992).

Baker, J. F. *The Steel Skeleton: Elastic Behaviour and Design*, Cambridge University Press (Cambridge, 1954).

Barr, A., and Feigenbaum, E. (eds.), *Handbook of Artificial Intelligence*, 3 volumes, Morgan Kaufmann (Los Altos, Calif., 1981).

Barraclough, K. C., *Steelmaking before Bessemer: Vol. 2, Crucible Steel*, Metals Society (London, 1984).

Barraclough, K. C., "Benjamin Huntsman, 1704–1776," Sheffield City Libraries, Local Studies Booklet (Sheffield, U.K., 1976).

Bartlett, F. C., *Remembering: A Study in Experimental and Social Psychology*, Cambridge University Press (Cambridge, 1932).

Basalla, G., *The Evolution of Technology*, Cambridge University Press (Cambridge, 1988).

Bell, C. G., and Newell, A., *Computer Structures: Readings and Examples*, McGraw-Hill (New York, 1971).

Berry, J. R., "Clifford E. Berry, 1918–1963: His Role in Early Computers," *Annals of the History of Computing*, 8, no. 4 (Oct. 1986), 361–389.

Billington, D. P., *Robert Maillart's Bridges: The Art of Engineering*, Princeton University Press (Princeton, N.J., 1979).

Billington, D. P., *The Tower and the Bridge*, Basic Books (New York, 1983).

Boden, M., *Artificial Intelligence and Psychology*, MIT Press (Cambridge, Mass., 1989).

Boden, M., *The Creative Mind*, Basic Books (New York, 1991).

Brachman, R., and Levesque, H. (eds.), *Readings in Knowledge Representation*, Morgan Kaufmann (Los Altos, Calif., 1985).

Broadbent, G., *Design in Architecture*, John Wiley & Sons (Chichester, U.K., 1973).

Brooks, F. P. Jr., *The Mythical Man-Month: Essays in Software Engineering*, Addison-Wesley (Reading, Mass., 1975).

Brown, D. C., and Chandrasekaran, B., *Design Problem Solving*, Pitman (London, 1989).

Burks, A. W., "From ENIAC to the Stored-Program Computer: Two Revolutions in Computers," in N. Metropolis, J. Howlett, and G.-C. Rota (eds.), *A History of Computing in the Twentieth Century*, Academic Press (New York, 1980).

Burks, A. W., and Burks, A. R., "The ENIAC: First General Purpose Electronic Computer," *Annals of the History of Computing*, 3, 4, Oct. 1981, pp. 310–399.

Bush, V., "The Differential Analyzer, A New Machine for Solving Differential Equations," *J. Franklin Institute*, 212 (1931), pp. 447–488.

Butterfield, H., *The Origins of Modern Science: 1300–1800*, Clarke, Irwin and Co. (Toronto, 1968).

Cahn, R. W., "Modern Practice in the Design of Strong Alloys," *Recent Developments in Metallurgical Science and Technology*, Vol. 4, Indian Institute of Metals, Silver Jubilee Symposium (New Delhi, 1972), pp. 527–551.

Cahn, R. W., "Recovery and Recrystallization" in R. W. Cahn (ed.), *Physical Metallurgy*, North Holland/Elsevier (Amsterdam and New York, 1970), pp. 1129–1197.

Campbell, D. T., "Blind Variation and Selective Retention in Creative Thoughts as in Other Knowledge Processes," *Psychological Review*, 67 (1960), 380–400.

Campbell, D. T., "Evolutionary Epistemology" in P. A. Schlip (ed.) *The Philosophy of Karl Popper*, Open Court Press (LaSalle, Ill., 1974).

Cardwell, D. S. L., *Turning Points in Western Technology*, Science History Publications (New York, 1972).

Cardwell, D. S. L., *The Fontana History of Technology*, Fontana Press (London, 1994).

Carlson, W. B., and Gorman, M. E., "A Cognitive Framework to Understand Technological Creativity: Bell, Edison and the Telephone," in R. J. Weber and D. N.

Perkins (eds.), *Inventive Minds: Creativity in Technology*, Oxford University Press (New York, 1992), pp. 48–79.

Chandrasekaran, B., "Design Problem Solving: A Task Analysis," *AI Magazine*, 11 (1990), 59–71.

Chandrasekaran, B., and Josephson, S. G., "Architecture of Intelligence: The Problem and Current Approaches to Solutions," *Current Science*, 64, no. 6 (March 25, 1993).

Chandrasekar, S., *Truth and Beauty*, University of Chicago Press (Chicago, 1987).

Charniak, E., and McDermott, D., *Introduction to Artificial Intelligence*, Addison-Wesley (Reading, Mass., 1985).

Chomsky, N., *Syntactic Structures*, Mouton (The Hague, 1969).

Chrimes, M., *Civil Engineering 1839–1889*. Alan Sutton/Thomas Telford (Shroud, U.K./London, 1991).

Clark, E., *The Britannia and Conway Tubular Bridges*, 2 volumes, Day & Sons (London, 1850).

Cohen, I. B., *The Newtonian Revolution*, Cambridge University Press (Cambridge, 1980).

Cohen, I. B., *Revolution in Science*, Harvard University Press (Cambridge, Mass., 1985).

Collins, A. M., and Loftus, E. F., "A Spreading-Activation Theory of Semantic Processing," *Psychological Reviews*, 82 (1975), 407–428.

Constant, E. W. II, *Origins of the Turbojet Revolution*, Johns Hopkins University Press (Baltimore, 1980).

Cotterall, B., and Kamminga, J., *Mechanics of Pre-Industrial Technology*, Cambridge University Press (Cambridge, 1990).

Cowan, H. J., *An Historical Outline of the Architectural Sciences*, Elsevier (New York, 1977).

Coyne, R. D., Rosenman, M. A., Radford, A. D., Balachandran, M., and Gero, J. S., *Knowledge Based Design Systems*, Addison-Wesley (Reading, Mass., 1989).

Craik, K., *The Nature of Explanation*, Cambridge University Press (Cambridge, 1942).

Crouch, T. D., "Why Wilbur and Orville? Some Thoughts on the Wright Brothers and the Process of Invention," in R. J. Weber and D. N. Perkins, *Inventive Minds*, Oxford University Press (New York, 1992).

Darling, A. S., "Non-Ferrous Metals," in I. McNeil (ed.), *An Encyclopedia of the History of Technology*, Routledge (London, 1990), pp. 47–145.

Das-Gupta, H. N., "The Development, Progress and Scientific Aspects of Powder Metallurgy," Presidential Address, Section of Engineering and Metallurgy, 48th Indian Science Congress, Roorkee, India, 1961.

Dasgupta, S., "The Organization of Microprogram Stores," *ACM Computing Surveys*, 11, no. 1 (Mar. 1979), 39–65.

Dasgupta, S., *Design Theory and Computer Science*, Cambridge University Press (Cambridge, 1991).

Dasgupta, S., "Two Laws of Design," *Intelligent Systems Engineering*, 1, no. 2 (Winter 1992), 146–156.

Dasgupta, S., *Creativity in Invention and Design*, Cambridge University Press (New York, 1994).

Dasgupta, S., "Testing the Hypothesis Law of Design: The Case of the Britannia Bridge," *Research in Engineering Design*, 6, no. 1 (1994), 38–57.

Dennett, D. C., *Brainstorms*, MIT Press (Cambridge, Mass., 1978).

Dennett, D. C., *The Intentional Stance*, MIT Press (Cambridge, Mass., 1983).

Dickinson, H. W., *A Short History of the Steam Engine*, Cambridge University Press (Cambridge, 1939).

Dijkstra, E. W., "Notes on Structured Programming," in O.-J. Dahl, E. W. Dijkstra, and C. A. R. Hoare, *Structured Programming*, Academic Press (New York, 1972).

Dijkstra, E. W., *A Discipline of Programming*, Prentice-Hall (Englewood Cliffs, N.J., 1976).

Dreyfus, H., *What Computers Can't Do*, Harper & Row (New York, 1979).

Eckert, J. P., "The ENIAC," in N. Metropolis, J. Howlett, and G.-C. Rota (eds.), *A History of Computing in the Twentieth Century*, Academic Press (New York, 1980), pp. 525–539.

Fairbairn, W., *An Account of the Construction of the Britannia and Conway Tubular Bridges*, John Weale/Longman, Brown, Green and Longmans (London, 1849).

Ferguson, E., *Engineering and the Mind's Eye*, MIT Press (Cambridge, Mass., 1992).

Fetzer, J., "Program Verification: The Very Idea," *Communications of the ACM*, 31, no. 9 (Sept. 1988), 1048–1063.

Findler, N. V. (ed.), *Associative Network: Representation and Use of Knowledge by Computers*, Academic Press (New York, 1974).

Floyd, R. W., "Assigning Meaning to Programs," *Mathematical Aspects of Computer Science*, American Mathematical Society (Providence, R.I., 1967).

Fodor, J. A., and Pylyshyn, Z. A., "Connectionism and Cognitive Architecture: A Critical Analysis," in S. Pinker and S. Mehler (eds.) *Connections and Symbols*, MIT Press (Cambridge, Mass., 1988), pp. 30–72.

Freeman, P., *Software Perspectives*, Addison-Wesley (Reading, Mass., 1987).

Gale, W. K. V., "Ferrous Metals" in I. McNeil (ed.), *An Encyclopedia of the History of Technology*, Routledge (London, 1990), pp. 157–168.

Galileo, *Dialogue Concerning Two New Sciences* (1638), translated by H. Crew and A. deSalvio, Dover (New York, 1954).

Gardner, J., *The Art of Fiction*, Vintage Books (New York, 1991).

Ghiselin, B., *The Creative Process: A Symposium*, University of California Press (Berkeley, 1952).

Gjertsen, D., *Science and Philosophy*, Penguin Books (London, 1989).

Goldstine, H. H., *The Computer from Pascal to von Neumann*, Princeton University Press (Princeton, N.J., 1972).

Goldstine, H. H., and Goldstine, A., "The Electronic Numeral Integrator and Computer (ENIAC)," *Mathematical Tables and Automatic Computation*, 2, no. 15 (1946), 97–110.

Gombrich, E. H., *Art and Illusion: A Study in the Psychology of Pictorial Representation*, Phaidon Books (London, 1960).

Gould, S. J., *Ontogeny and Phylogeny*, Belknap Press of Harvard University Press (Cambridge, Mass., 1977).

Gregory, R. L., *Mind in Science*, Weidenfeld and Nicolson (London, 1981).

Gries, D., *The Science of Programming*, Springer-Verlag (New York, 1981).

Groak, S., *The Idea of Building*, E. & F. N. Spon (London, 1992).

Gruber, H. E., *Darwin on Man: A Psychological Study of Scientific Creativity*, 2nd edition, University of Chicago Press (Chicago 1981).

Gruber, H. E., Terrell, G., and Wertheimer, M. (eds.), *Contemporary Approaches to Creative Thinking*, Atherton Press (New York, 1963).

Habib, S., "A Brief Chronology of Microprogramming Activity" in S. Habib (ed.)

Microprogramming and Firmware Engineering Methods, van Nostrand-Reinhold (New York, 1988), pp. 1–32.

Hadamard, J., *The Psychology of Invention in the Mathematical Field*, Princeton University Press (Princeton, N.J., 1945).

Harding, R., *An Anatomy of Inspiration*, W. Heffer & Sons (Cambridge, 1942).

Hardy, G. H., *Ramanujan*, Cambridge University Press (Cambridge, 1940).

Hesse, M. B., *Models and Analogies in Science*, Sheed and Ward (London, 1966).

Hoare, C. A. R., "An Axiomatic Approach to Computer Programming," *Communications of the ACM*, 12, 10, Oct., 1969, pp. 576–580, 583.

Hoare, C. A. R., *The Mathematics of Programming*, Inaugural Lecture, University of Oxford, Clarendon Press (Oxford, 1986).

Hobsbawm, E. J., *The Age of Revolution*, Weidenfeld and Nicolson (London, 1962).

Hodges, A., *Alan Turing: The Enigma*, Simon and Schuster (New York, 1983).

Holland, J. H., Holyak, K. J., Nisbett, R. E., and Thagard, P. R., *Induction*, MIT Press (Cambridge, Mass., 1986).

Holton, G., *An Introduction to Concepts and Theories in Physical Sciences*, Addison-Wesley (Reading, Mass., 1952).

Holton, G., and Elkana, Y. (eds.), *Albert Einstein: Historical and Cultural Perspectives*, Princeton University Press, (Princeton, N.J. 1982).

Hulme, E. W., "The Pedigree and Career of Benjamin Huntsman, Inventor in Europe of Crucible Steel," *Transactions of the Newcomen Society*, 24, (1994/95), 37–48.

Husson, S. S., *Microprogramming: Principles and Practices*, Prentice-Hall (Englewood Cliffs, N.J., 1970).

James, C. L. R., *Beyond a Boundary*, Stanley Paul (London, 1963).

Jeffrey, L., "Writing and Rewriting Poetry: William Wordsworth," in D. B. Wallace and H. E. Gruber, *Creative People at Work*, Oxford University Press (New York, 1989), pp. 69–89.

Johnson-Laird, P. N., *The Computer and the Mind*, Harvard University Press (Cambridge, Mass., 1988).

Jones, J. C., *Design Methods: Seeds of Human Future*, John Wiley & Sons (New York, 1980).

Jones, J. C., and Thornley, D. G. (eds.), *Conference on Design Methods*, Pergamon Press and Macmillan (Oxford and New York, 1963).

Jurgen, R. K., "Jacob Rabinow," *IEEE Spectrum*, Dec. 1991, 24–25.

Kanigel, R., *The Man Who Knew Infinity*, Charles Scribner's Sons (New York, 1991).

Kant, E., "Understanding and Automating Algorithm Design." *IEEE Transactions on Software Engineering*, SE-11, no. 11 (Nov. 1985), 1361–1374.

Kant, E., and Newell, A., "Problem Solving Techniques for the Design of Algorithms," *Information Processing and Management*, 20, nos. 1–2 (1986), 97–118.

Knuth, D. E., *The Art of Computer Programming*, Addison-Wesley (Reading, Mass.), *Volume 1: Fundamental Algorithms* (1968; 2nd edition 1973); *Volume 2: Seminumerical Algorithms* (1969; 2nd edition 1981); *Volume 3: Sorting and Searching* (1973).

Knuth, D. E., "Computer Programming as an Art," *Communications of the ACM*, 17, 12 (Dec. 1974), 667–673.

Knuth, D. E., *Computers and Typesetting*, 5 volumes, Addison-Wesley (Reading, Mass., 1986).

Knuth, D. E., "The Error of TEX," *Software: Practice and Experience*, 19, no. 7 (1989), 607–685.

Knuth, D. E., and Floyd, R. W., "Notes on Avoiding 'Go To' Statements," *Informa- tion Processing Letters*, 1 (1971), 21–31.

Koestler, A., *The Act of Creation*, Hutchinson & Co. (London, 1964).

Kuhn, T. S., *The Copernican Revolution*, Harvard University Press (Cambridge, Mass., 1957).

Kuhn, T. S., *The Structure of Scientific Revolutions*, 2nd edition, University of Chi- cago Press (Chicago, 1970; 1st edition 1962).

Kuhn, T. S., "Logic of Discovery or Psychology of Research," in I. Lakatos and A. Musgrave, *Criticism and the Growth of Knowledge*, Cambridge University Press (Cambridge, 1970), pp. 1–24.

Kulkarni, D., and Simon, H. A., "The Processes of Scientific Discovery: The Strategy of Experimentation," *Cognitive Science*, 12 (1988), 139–176.

Lachman, R., Lachman, J. L., and Butterfield, E. C., *Cognitive Psychology and Infor- mation Processing*, Lawrence Erlbaum Associates (Hillside, N.J., 1979).

Lakatos, I., "Falsification and the Methodology of Scientific Research Programmes," in I. Lakatos and A. Musgrave (eds.), *Criticism and the Growth of Knowledge*, Cambridge University Press (Cambridge, 1970), pp. 91–196.

Langley, P., Simon, H. A., Bradshaw, G. L., and Zytkow, J. M., *Scientific Discovery*, MIT Press (Cambridge, Mass., 1987).

Laudan, L., *Progress and Its Problems*, University of California Press (Berkeley, 1977).

Laudan, L., *Science and Values*, University of California Press (Berkeley, 1984).

Lavington, S. H., "Computer Developments at Manchester University," in N. Metropo- lis, J. Howlett, and G.-C. Rota (eds.), *A History of Computing in the Twenti- eth Century*, Academic Press (New York, 1980), pp. 433–443.

Layton, E. T., Jr., "Mirror-Image Twins: The Communities of Science and Technology in 19th Century America," *Technology and Culture*, 12, no. 4 (Oct. 1971), 562–580.

Layton, E. T., Jr., "Technology as Knowledge," *Technology and Culture*, 15, no. 1 (Jan. 1974), 31–41.

Layton, E. T., Jr., "American Ideologies of Science and Engineering," *Technology and Culture*, 17, no. 4 (Oct. 1976), 688–701.

Lowes, J. L., *The Road to Xanadu*, Houghton Mifflin (Boston, Mass., 1927).

Mauchly, J. W., "The ENIAC," in N. Metropolis, J. Howlett, and G.-C. Rota (eds.), *A History of Computing in the Twentieth Century*, Academic Press (New York, 1980), pp. 541–550.

Mauchly, J. W., "The Use of High Speed Vacuum Tubes for Calculating," unpublished memorandum, University of Pennsylvania, Philadelphia, Aug. 1942. Reprinted in B. Randall (ed.), *The Origins of Digital Computers, Selected Papers*, 2nd edition, Springer-Verlag (New York, 1975), pp. 329–332.

Mauchly, K. K., "John Mauchly's Early Years," *Annals of the History of Computing*, 6, no. 2 (Apr. 1984), 116–138.

Maynard Smith, J., *On Evolution*, Edinburgh University Press (Edinburgh, 1972).

Mayr, E., *Towards a New Philosophy of Biology*, Belknap Press of Harvard Univer- sity Press (Cambridge, Mass., 1988).

McGannon, H. E. (ed.), *The Making, Shaping and Treating of Steel*, 9th edition, United States Steel (Pittsburgh, 1970).

McNeil, I., "Roads, Bridges and Vehicles" in I. McNeil (ed.), *An Encyclopedia of the History of Technology*, Routledge (London and New York, 1990), pp. 431–473.

Medawar, P. B., *The Art of the Soluble*, Penguin Books (Harmondsworth, U.K., 1969).

Medawar, P. B., "Is the Scientific Paper a Fraud?," in P. B. Medawar, *The Threat and the Glory: Reflections on Science and Scientists*, Oxford University Press (Oxford, 1990), pp. 228–233.

Medawar, P. B., and Medawar, J. S., *Aristotle to Zoos: A Philosophical Dictionary of Biology*, Harvard University Press (Cambridge, Mass., 1983).

Mill, J. S., *A System of Logic: Ratiocinative and Inductive* (1843; reprint, University of Toronto Press [Toronto 1974]).

Miller, G. A., "The Magical Number Seven, Plus or Minus Two: Some Limits on Our Capacity for Information Processing," *Psychological Review*, 63 (1956), 81–97.

Minsky, M., *Computation: Finite and Infinite Machines*, Prentice-Hall (Englewood Cliffs, N.J., 1967).

Minsky, M., "A Framework for Representing Knowledge" in P. Winston (ed.), *The Psychology of Computer Vision*, McGraw-Hill (New York, 1975), pp. 211–277.

Mollenhoff, C. R., *Atanasoff: Forgotten Father of the Computer*, Iowa State University Press (Ames, 1988).

Mumford, L., *Technics and Civilization*, Harcourt, Brace & World Inc. (New York, 1963).

Nervi, P. L., *The Works of Pier Luigi Nervi*, F. A. Praeger (New York, 1957).

Nervi, P. L., *Aesthetics and Technology in Building*, Harvard University Press (Cambridge, Mass., 1966).

Newell, A., "The Knowledge Level," *Artificial Intelligence*, 18 (1982), 87–127.

Newell, A., *Unified Theories of Cognition*, Harvard University Press (Cambridge, Mass., 1990).

Newell, A., Rosenbloom, P. S., and Laird, J., "Symbolic Architectures for Cognition," in M. I. Posner (ed.), *Foundations of Cognitive Science*, MIT Press (Cambridge, Mass., 1989), pp. 93–131.

Newell, A., Shaw, J. C., and Simon, H. A., "The Process of Creative Thinking," in H. E. Gruber, G. Terrell, and M. Wertheimer (eds.), *Contemporary Approaches to Creative Thinking*, Atherton Press (New York, 1963), pp. 63–119.

Newell, A., and Simon, H. A., "Elements of a Theory of Human Problem Solving," *Psychological Review*, 65 (1958), 151–166.

Newell, A., and Simon, H. A., *Human Problem Solving*, Prentice-Hall (Englewood Cliffs, N. J., 1972).

Norman, D. A., *The Design of Everyday Things*, Doubleday (New York, 1989).

Pais, A., *'Subtle is the Lord . . .': The Science and Life of Albert Einstein*, Oxford University Press (Oxford, 1982).

Patrick, C., "Creative Thoughts in Poets," in R. Woodworth (ed.), *Archives of Psychology*, 178 (1935).

Patrick, C., "Creative Thoughts in Artistic Activity," *Journal of Psychology*, 4 (1937), 35–73.

Penrose, R., *The Emperor's New Mind*, Oxford University Press (Oxford, 1989).

Petroski, H., *To Engineer Is Human: The Role of Failure in Successful Design*, St. Martin's Press (New York, 1985).

Petroski, H., *The Pencil: A History of Design and Circumstance*, Alfred A. Knopf (New York, 1989).

Petroski, H., "Paconius and the Pedestal for Apollo: A Case Study of Error in Conceptual Design," *Research in Engineering Design*, 3 (1991), 123–128.

Petroski, H., "Vitruvius's Auger and Galileo's Bones: Paradigm of Limit to Size in Design," *Journal of Mechanical Design*, 114 (Mar. 1992), 23–28.

Petroski, H., *The Evolution of Useful Things*, Alfred A. Knopf (New York, 1992).

Petroski, H., "The Britannia Tubular Bridge," *American Scientist*, May–June 1992, 220–224.

Petroski, H., "Galileo's Confirmation of a False Hypothesis: A Paradigm for Logical Error in Design," *Civil Engineering System*, 9 (1992), 251–263.

Petroski, H., *Design Paradigms: Case Histories of Error and Judgement in Engineering*, Cambridge University Press (New York, 1994).

Poincaré, H., *The Foundations of Science*, translated by G. B. Halstead (1908; reprint, Science Press [New York 1952]).

Polanyi, M., *Personal Knowledge*, University of Chicago Press (Chicago, 1962).

Polanyi, M., *Knowing and Being*, University of Chicago Press (Chicago, 1969).

Polya, G., *How to Solve It*, 2nd edition, Princeton University Press (Princeton, N.J., 1957).

Popper, K. R., *The Logic of Scientific Discovery*, Harper & Row (New York, 1968).

Popper, K. R., *Objective Knowledge*, Clarendon Press (Oxford, 1972).

Posner, M. I. (ed.), *Foundations of Cognitive Science*, MIT Press (Cambridge, Mass., 1989).

Pye, D., *The Nature of Design*, Studio Vista and Reinhold Publishing Corporation (London and New York, 1964).

Pye, D., *The Nature and Aesthetics of Design*, Herbert Press (London, 1978).

Pylyshyn, Z., "Computation and Cognition: Issues in the Foundation of Cognitive Science," *Behavioral and Brain Sciences*, 3, no. 1 (1980), 154–169.

Pylyshyn, Z., *Computation and Cognition*, MIT Press (Cambridge, Mass., 1984).

Randell, B., "Origins of Digital Computers: Supplementary Bibliography," in N. Metropolis, J. Howlett, and G.-C. Rota, (eds.), *A History of Computing in the Twentieth Century*, Academic Press (New York, 1980), pp. 629–659.

Reason, J., *Human Error*, Cambridge University Press (Cambridge, 1990).

Reed-Hill, R. E., *Physical Metallurgy Principles*, 2nd edition, D. Van Nostrand Company (New York, 1973).

Reiter, R., "Nonmonotonic Reasoning," *Annual Review of Computer Science*, 2 (1987), 147–186.

Reitman, W., *Cognition and Thought*, John Wiley (New York, 1965).

Rolt, L. T. C., *Thomas Newcomen*, David and Charles/Dawlish MacDonald (London, 1963).

Rosenberg, N., and Vincenti, W. G., *The Britannia Bridge: The Generation and Diffusion of Technological Knowledge*, MIT Press (Cambridge, Mass., 1978).

Rumelhart, D. E., and McClelland, J. L., *Parallel Distributed Processing, Vol. 1: Foundations*, MIT Press (Cambridge, Mass., 1986).

Ruse, M., *Taking Darwin Seriously*, Basil Blackwell (Oxford, 1986).

Rychener, M. D. (ed.), *Expert Systems for Engineering Design*, Academic Press (New York, 1988).

Schank, R. C., and Abelson, R. P., *Scripts, Plans, Goals and Understanding*, Lawrence Erlbaum (Hillsdale, N.J., 1977).

Schön, D., *The Reflective Practitioner*, Basic Books (New York, 1983).

Searle, J., *Intentionality: An Essay in the Philosophy of Mind*, Cambridge University Press (Cambridge, 1983).

Searle, J., *Mind, Brain and Science*, Harvard University Press (Cambridge, Mass., 1984).

Searle, J., *The Rediscovery of Mind*, MIT Press (Cambridge, Mass., 1992).

Shannon, C. E., "A Symbolic Analysis of Relay and Switching Circuits," *American Institute of Electrical Engineers Transactions*, 57 (1938), pp. 713–723.

Shaw, M., "Prospects for an Engineering Discipline of Software," *IEEE Software*, Nov. 1990, pp. 15–24.

Shrager, J., and Langley, P. (eds.), *Computational Models of Scientific Discovery and Theory Formation*, Morgan Kaufman (San Mateo, Calif., 1990).

Simon, H. A., "The Architecture of Complexity," *Proc. Amer. Phil. Society*, 106 (Dec. 1962), 467–482.

Simon, H. A., "Theories of Bounded Rationality," in C. B. Radner and R. Radner (eds.), *Decision and Organization*, North-Holland (Amsterdam, 1972), pp. 161–176.

Simon, H. A., "The Structure of Ill-Structured Problems," *Artificial Intelligence*, 4 (1973), 181–200.

Simon, H. A., *Administrative Behavior*, 3rd edition, Free Press (New York, 1976; 1st edition, 1946).

Simon, H. A., *The Sciences of the Artificial*, 2nd edition, MIT Press (Cambridge, Mass., 1981; 1st edition 1969).

Simon, H. A., *Models of Bounded Rationality*, 2 volumes, MIT Press (Cambridge, Mass., 1982).

Simon, H. A., *Reason in Human Affairs*, Basil Blackwell (Oxford, 1983).

Simonton, D. K., *Scientific Genius: A Psychology of Science*, Cambridge University Press (Cambridge, 1988).

Smiles, S., *Industrial Biography: Iron Workers and Tool Makers*, John Murray (London, 1863; reprinted with Introduction by L. T. C. Rolt, David & Charles [Newton Abbot, U.K., 1967]).

Smith, C. S., "The Discovery of Carbon in Steel," *Technology and Culture*, 5 (1965), 149–175.

Smith, C. S., "Matter versus Material: A Historical View," *Science*, 162 (1968), 637–644.

Smith, C. S., *A Search for Structure*, MIT Press (Cambridge, Mass., 1982).

Somerscales, E. F. C., "Steam and Internal Combustion Engines," in I. McNeil (ed.), *An Encyclopedia of the History of Technology*, Routledge (London, 1980), pp. 272–349.

Sommerville, I., *Software Engineering*, 3rd edition, Addison-Wesley (Reading, Mass., 1989).

Sperry, R., *Science and Moral Priority*, Basil Blackwell (Oxford, 1983).

Steadman, J. P., *The Evolution of Designs*, Cambridge University Press (Cambridge, 1979).

Sternberg, R. J., "A Three-Facet Model of Creativity," in R. J. Sternberg (ed.), *The Nature of Creativity*, Cambridge University Press (Cambridge, 1988), pp. 125–147.

Sturt, G., *The Wheelwright's Shop*, Cambridge University Press (Cambridge, 1923; Canto Edition, 1993).

Thagard, P., *Computational Philosophy of Science*, MIT Press (Cambridge, Mass., 1988).

Thagard, P., "The Conceptual Structure of the Chemical Revolution," *Philosophy of Science*, 57 (1990), 183–209.

Thagard, P., *Conceptual Revolutions*, Princeton University Press (Princeton, N.J., 1992).

Thagard, P., and Nowak, G., "The Conceptual Structure of the Geological Revolution," in J. Shrager and P. Langley (eds.), *Computational Models of Scientific Discovery and Theory Formation*, Morgan Kaufman (San Mateo, Calif., 1990).

Thurston, R. H., *A History of the Growth of the Steam Engine*, Cornell University Press (Ithaca, N.Y., 1939; 1st edition 1878).

Timoshenko, S., *History of Strength of Materials*, McGraw-Hill (New York, 1953; reprint, Dover [New York, 1983]).

Turing, A. M., "On Computable Numbers with an Application to the Eintscheidungs-problem," *Proc. London Mathematical Society*, series 2, 42 (1936), 230–265.

Turing, A. M., "Computing Machinery and Intelligence," *Mind*, 59 (Oct. 1950), 430–460.

Usher, A. P., *A History of Mechanical Inventions*, revised edition 1954, Harvard University Press (Cambridge, Mass., 1954; reprint, Dover [New York, 1988]; 1st edition 1929).

Vincenti, W. G., *What Engineers Know and How They Know It*, Johns Hopkins University Press (Baltimore, 1992).

Vincenti, W. G., "Engineering Knowledge, Type of Design, and Level of Hierarchy: Further Thoughts About *What Engineers Know . . .*" in P. Kroes and M. Bakker (eds.), *Technological Development and Science in the Industrial Age*, Kluwer Academic Publishers (Boston, 1992).

Vitruvius, *De Architectura* (circa 25 B.C.), translated by M. H. Morgan as *Ten Books on Architecture*, Harvard University Press (Cambridge, Mass., 1914; reprint, Dover [New York, 1960]).

von Neumann, J., "First Draft of a Report on the EDVAC," Moore School of Electrical Engineering, University of Pennsylvania (Philadelphia, 1945). Reprinted in B. Randell (ed.), *The Origins of Digital Computers: Selected Papers*, Springer-Verlag (New York, 1975), pp. 355–365.

Wallace, D. B., and Gruber, H. E. (eds.), *Creative People at Work*, Oxford University Press (New York, 1989).

Wallas, G., *The Art of Thought*, Harcourt, Brace, Jovanovich (New York, 1926).

Watson, G. F., "Mesaru Ibuka," *IEEE Spectrum*, Dec. 1991, pp. 22–28.

Weber, R. J., "Stone Age Knife to Swiss Army Knife: An Invention Prototype," in R. J. Weber and D. N. Perkins (eds.), *Inventive Minds: Creativity in Technology*, Oxford University Press (New York, 1992), pp. 217–237.

Weisberg, R. W., *Creativity: Genius and Other Myths*, W. H. Freeman (New York, 1986).

Weisberg, R. W., *Creativity: Beyond the Myth of Genius*, W. H. Freeman (New York, 1993).

Wilkes, M. V., "The Best Way to Design an Automatic Calculating Machine," *Report, Manchester University Computer Inaugural Conference*, Manchester, U.K., July 1951. Reprinted in *Annals of the History of Computing* 8, no. 2 (Apr. 1986), 116–118.

Wilkes, M. V., *Automatic Digital Computers*, Methuen & Co. (London, 1957).

Wilkes, M. V., "The Growth of Interest in Microprogramming: A Literature Survey," *ACM Computing Surveys*, 1, no. 3 (1969), 139–145.

Wilkes, M. V., "Babbage as a Computer Pioneer," *Historia Mathematica*, 4 (1977), 415–440.

Wilkes, M. V., *Memoirs of a Computer Pioneer*, MIT Press (Cambridge, Mass., 1985).

Wilkes, M. V., "The Genesis of Microprogramming," *Annals of the History of Computing*, 8, no. 2 (Apr. 1986), 116–118.

Winston, P. A., *Artificial Intelligence*, 2nd edition Addison-Wesley (Reading, Mass., 1984).

Wirth, N., "Program Development by Stepwise Refinement," *Communications of the ACM*, 14, no. 4 (Apr. 1971), 221–227.

Wolpert, L., *The Unnatural Nature of Science*, Faber and Faber (London, 1992).

Index

Abacus, 163
Abduction, 108, 109, 110, 205n
Abelson, R. P., 196n
Accumulator, 136, 141, 208n
Action, deliberative, 30
  knowledge level, 32, 38ff, 42, 48, 87, 108, 113, 159, 181
  non-rational, 43, 48
  rational, 43, 48
Add-subtract mechanism, 162, 163
Addis, William, 14, 24, 25, 50, 59, 187n, 189n, 192n, 196n, 198n, 201n, 211n
Age hardening, 70, 71, 87, 98, 99, 152, 200n
Age of Automation, 134
Agent, cognitive, 31
Agricola, Georgius, 35, 194n
Aiken, Howard, 149, 162, 210n
Air pump, 126
Airplane, invention of, 59, 92
Alexander, Christopher, 5, 12, 14, 35, 50, 187n, 188n, 194n
Algorithm, 36, 69
Algorithm, design of, 69, 74ff, 100
Alloys
  as artifacts, 153
  corrosion resistant, 69, 97, 98, 152
  creep resistant, 21, 69ff, 122, 152, 153
  for gas turbine blades, 21, 69ff
  heat treatment of, 4, 168
  nickel-chromium, 71ff, 87, 97, 121, 122, 152, 155, 171, 183
  phase transformations in, 4, 168
Analog computing, 210n
Analogical reasoning, 80, 201n
Analytical Engine, 8, 16, 148, 210n
Anderson, J. R., 195n
Arbib, Michael, 196n
Arch design, 175, 176
Archimedes, xiii

Artifact
  abstract, 11, 12, 16, 23, 189n
  definition of, 9
  evolution of, 9
  fallibility of, 17ff
  making of, 3
  manufacturing process as, 14, 15
  material, 10ff, 16
Artificial intelligence (AI), 31, 102, 150, 167, 193n
Aspray, William, 195n
Atanasoff, John, 54, 137, 138, 139, 140, 147, 149, 162, 208n, 210n, 212n
Atanasoff-Berry Computer (ABC), 54, 138, 140, 147, 149, 162, 163, 210n
Atkinson, R. C., 188
Atmospheric pressure, 126, 127, 161, 207n
Atmospheric steam engine, 54, 124ff, 131, 134, 137, 149, 160, 161, 185, 207n
Authement, Ray, xv
Automata theory, 60

Babbage, Charles, 8, 16, 148, 149, 162, 189n, 210n
Bagchi, Amiya, xvii
Baker, John F., 25, 192n
Ballistic Research Laboratory, 22, 137
Bandwidth, 118, 119, 195n, 206n
Barr, A., 193n, 195n
Barraclough, Kenneth, 62, 191n, 192n, 198n, 199n, 206n
Bartlett, Frederick, 196n
Basalla, George, 123, 148, 187n, 188n, 197n, 198n, 206n, 210n
Beam design, 80, 82, 157, 173
Bell Telephone Laboratories, 142, 147
Bell, Alexander Graham, 3, 24, 27, 54, 55, 91, 121
Bell, C. Gordon, 29, 192n
Bennett, Albert, 145

Bennett, Arnold, 13
Berry, Clifford, 54, 138, 139, 147
Berry, J. R., 208n
Bessemer converter, 46
Bessemer process, 53
Bessemer, Henry, 53, 62
Billington, David, 38, 179, 194n, 199n, 214n
Biological level of cognition, 30, 31
Bisociation, 49, 91ff, 101, 114, 121
Black's law, 186
Blast furnace, 4, 122
Boden, Margaret, 55, 193n, 197n
Bollee's multiplication algorithm, 142ff, 147
Boolean algebra, 163
Borrowing of ideas, 114
Bounded rationality, 43, 45, 48, 66, 67, 90, 116, 118, 182, 195n, 199n
Boyle's law, 55, 186
Boyle, Robert, 55, 126
Brachman, R., 195n
Bradshaw, G. L., 204n
Brainerd, J. G., 22, 140, 145
Brass founding, 121
Bridges
    cast iron type, 79, 100, 115
    reinforced concrete type, 179
    suspension type, 79, 100, 110, 114, 115, 158, 210n
    tubular type, 80, 101, 110, 122, 158
    truss type, 80, 101, 114, 210n
    wrought iron type, 80, 82, 101, 110, 157
Britannia Bridge, 13, 17, 18, 21, 28, 33, 52, 63, 78ff, 85, 86, 88, 100ff, 110, 116, 121, 149, 157, 158, 171, 174, 177, 199n, 200n
Broadbent, Geoffrey, 194n
Brooks, Frederick, 165, 212n
Brown, David C., 188n, 200n
Brunel, Isambard, 214n
Buckling of beams, 81, 82, 159, 177, 178
Burks, Alice, 137, 139, 142, 143, 145, 208n
Burks, Arthur, 22, 134, 137, 139, 140, 142, 143, 144, 145, 162, 190n, 191n, 197n, 208n
Bush, Vannevar, 137, 139, 147, 151, 208n
Butterfield, E. C., 189n
Butterfield, Herbert, 6, 187n

Cahn, R. W., 69, 97, 190n, 200n, 203n
Caldwell, Samuel, 139, 163, 212n
Calley, John, 124, 131
Campbell, Donald, 92, 95, 203n
Cardwell, Donald, xvii, 124, 198n, 206n, 207n

Carlson, Bernard, 24, 91, 103, 191n, 197n, 203n, 204n
Cech, Claude, xvii
Cementation process, 62
Chance-configuration theory, 93ff, 96
Chance permutation, 94
Chandrasekaran, B., 66, 188n, 193n, 195n, 200n
Chandrasekhar, S., 4, 46, 187n
Charcoal, 21
Charles II, King, 127
Charniak, E., 193n
Chemical revolution, 59, 195n
Chester and Holyhead Railway, 21, 79, 81, 84
Chicago Bulls, 46
Chomsky, Noam, 193n
Chrimes, M., 199n
Chrysler, Walter, 214n
Chunk, 14
Chunking, 47
Clark, Edwin, 13, 78, 82, 111, 160, 188n, 191n, 199n, 201n, 202n, 204n, 205n, 211n, 213n
Clarke, Thomas C., 151, 154
Clock, invention of, 59
Coalbrookdale iron works, 15
Cognition, 30, 31
Cognitive agent, 31
Cognitive psychology, 31, 102
Cognitive science, 31, 213n
Cognitive system, 29
Cognitive unconscious, 89
Cohen, I. Bernard, 198n, 206n
Coherency in crystal structures, 200n
Coleridge, Samuel Taylor, 89, 90
Collins, A. M., 195n
Communication configuration, 94, 95
Complexity, 29
Compression in beams, 159, 160
Computability, 60
Computer memory, 21, 22, 136
Computer programming, 15, 16, 36, 66, 91
Computer-aided design, 8
Computers
    binary type, 209n
    control unit of, 23, 26, 33, 54, 91, 104, 109
    electromechanical, 22, 149, 162
    electronic, 22, 134ff, 140, 148, 149, 162, 197n, 208n
    relay type, 142
    stored program, 15, 22, 23, 104, 109, 136, 148, 163, 204n
Conceptual matrix, 91

Conceptual model, 214n
Conceptual network, 195n
Conjecture and refutation, 87
Connectionism, 31
Conscious thought, 38, 89, 108, 120
Consolidated Aircraft Corporation, 96
Constant, Edward, 95, 148, 203n, 210n
Control volume technique, 160
Convex hull algorithm, 74ff, 88, 99
Conway Bridge, 199n
Copernican revolution, 59
Copernican theory, 37
Corrosion resistance, 69, 97, 98
Cotterell, Brian, 174, 213n
Cowan, H. J., 194n
Coyne, Richard, 188n
Craft, 9, 12ff
Craik, Kenneth, 214n
Cranege brothers, 15
Cranege process, 15
Crawford, Perry, 144
Creativity
    as heavenly inspiration, xiv
    historical, 56
    in technology, xiii, xiv
    in writing, 90, 206n
    its relation to orginality, 55ff
    psychological, 56
Creativity
    HN, 57, 102
    HO, 57, 61, 63, 65, 95, 102
    PN, 57, 102, 117
    PO, 57, 61, 65, 95, 102
Creep resistance, 69, 97, 98
Crick, F. H. C., 55
Cricket, the sport of, 188n
Crucible process, 24, 55, 62, 63, 121, 191n, 199n
Crouch, Tom, 92, 203n
Curiosity, 25

Dahl, O-J., 166, 194n
Dalton, John, 81
Darby, Abraham, 15, 190n
Darling, A. S., 211n
Dartnall, Terry, xvii
Darwin, Charles, 4, 7, 53, 55, 59, 92
Darwinian models of creativity, 92ff, 97, 123
Darwinian revolution, 59
Das-Gupta, H. N., 213n
Dasgupta, Subrata, 187n, 189n, 191n, 195n, 196n, 197n, 198n, 200n, 201n, 203n, 204n, 205n, 206n, 209n, 212n, 214n
Davis, David R., 96

Davis wing design, 96, 100
de Caus, Solomon, 127, 132, 160
Decidability problem, 60
Declarative knowledge, 33
Delhi iron pillar, 175
della Porta, Giambattista, 162
Dennett, Daniel, 31, 192n, 193n
Description levels, 29, 30
Design
    as conceptualization, 12ff, 50, 178
    computer-aided systems for, 8
    conceptual, 84
    creative act of, 6, 30, 181
    detailed, 84
    folk theories of, 66
    Gothic revolution in, 60
    Greek revolution in, 14, 59
    as the initiation of change, 51
    its distinction from making, 12ff, 178
    its relation to engineering science, 59, 60
    its relation to invention, 5, 53ff, 65, 182
    knowledge-based systems in, 8
    as a knowledge level process, 49ff
    normal, 52, 84, 95
    objectives of, 5, 50
    paradigms for, 66
    radical, 52
    science of, 5
    uniqueness of, 6
    unselfconscious process of, 12, 13
Design methods movement, 5
Design process, 30, 50, 66, 118, 119, 205n
    evolutionary nature of, 66
    universal characteristics of, 6, 50
Design space, 117
Design theory, 5
Dickinson, H. W., 127, 132, 206n, 207n, 212n
Difference Engine, 148
Differential analyzer, 22, 137, 138, 139, 140, 141, 145, 147, 149
Differential equations, solving of, 139
Dijkstra, Edgser, 36, 166, 189n, 194n
Diode matrix, 91, 105, 106, 121, 163, 164
Dissatisfaction, 20, 22ff, 26, 27
Doric style, 34
Dreyfus, H., 193n
Dudley, Dud, 21, 190n
Durfee, B. M., 210n

Eckert, Presper, 21, 22, 23, 53, 54, 134, 138, 140, 145, 147, 149, 185, 191n, 208n, 210n
Edison, Thomas Alva, 3

EDSAC, 23, 53, 104, 105, 106, 109, 121, 189n, 191n, 205n
EDSAC 2, 54, 61
Einstein, Albert, 38, 59, 95, 194n
Electrostatic storage tube, 21
Eliot, Thomas Stearns, 29
Elkana, Y., 194n
Emerson, Ralph Waldo, 13
ENIAC, 22, 23, 54, 134ff, 142, 149, 185, 190n, 191n, 204n, 208n, 209n
Environment
  inner, 10ff, 15, 16, 18
  outer, 10ff, 15, 16, 18
Evans, Oliver, 214n
Error, types of, 18, 213n
Evolution
  by natural selection, 53
  theory of, 7, 92, 93
Evolutionary epistemology, 92, 122

Fairbairn, William, 13, 18, 33, 64, 78, 81, 82, 83, 111, 157, 158, 160, 171, 188n, 199n, 201n, 202n, 211n
Falsificationism, 201
Feigenbaum, E., 193n, 195n
Ferguson, Eugene, 188n, 199n, 214n
Fermat's last theorem, 190n
Fetzer, James, 189n
Fiction, xiii
File system design, 118ff
Fire-making, invention of, 59
Floyd, Robert, 165, 189n, 212n
Flux, 194n
Fodor, Jerry, 193n
Ford, Henry, 122
Forrester, Jay, 162
Frames, 47, 50
Free body technique, 160
Freeman, Peter, 212n

Gale, W. K. V., 196n
Galilei, Galileo, 3, 6, 46, 126, 173, 187n, 211n
Gardner, John, 90, 202n, 206n
Gas turbine, 21, 69ff, 152
Gaussian elimination, method of, 138
Generalization, 107, 108, 110, 114
Geological revolution, 196n
German steel, 24, 192n
Gero, John, xvii
Ghiselin, Brewster, xiii, 13, 95, 188n, 202n, 203n
Ghoshal, T. K., xvii
Gjertsen, Derek, 190n
Goals, 4, 18, 27ff, 31, 39, 75
Goldstine, Adele, 208n

Goldstine, Herman, 22, 134, 140, 144, 208n
Gombrich, E. H., 20, 190n
Gorman, Michael, 24, 91, 103, 191n, 197n, 203n, 204n
Gothic cathedrals, 6, 17
Gothic design revolution, 60
Gould, Stephen Jay, 206n, 210n
Greek design revolution, 60
Gregory, Richard, 31, 193n
Gries, David, 189n
Groak, S., 194n
Gruber, Howard, 97, 203n
Guernica, 8
Gutenberg, Johan, 59

Habib, S., 198n
Hadamard, Jacques, 89, 91, 95, 202n
Haeckel, Ernst, 209n
Hamilton, F. E., 210n
Harding, Rosamund, xiii, 202n
Hardy, G. H., 57, 58, 60, 197n, 198n
Harvard Mark-I, 210n
Hawking, Stephen, 162
Henry, Joseph, 151, 154
Hesse, Mary, 196n, 201n
Heuristics, 34, 115, 167
Hierarchical relationship, 29, 156
High technology, 162
Historical novelty, 57
Historical originality, 57
History of technology, xiii, 7, 14, 62, 162
Hoare, C. A. R., 165, 166, 189n, 194n
Hobsbawm, E. J., 206n
Hodges, Andrew, 198n
Hodgkinson, Eaton, 81, 160, 171
Holland, John, 108, 195n, 201n, 205n
Holton, Gerald, 194n
Holyoak, K. J., 195n, 205n
Hooke, Robert, 133
Hopper, Grace Murray, 162, 210n
Howlett, J., 208n
Hulme, E. W., 192n
Huntsman, Benjamin, xvii, 24, 55, 62, 121, 192n
Husson, S. S., 198n
Huxley, Thomas Henry, 4
Huygens, Christiaan, 126, 132
Hypothesis, 33, 66, 97, 100, 114, 149, 171
Hypothesis law of maturation, 66ff, 87, 96, 97, 116, 184, 200n

IBM crosspoint multiplier, 143, 144, 147
IBM plugboard concept, 144, 149
IBM punched card equipment, 142, 147, 210n
IBM System/360 computer, 61

Ibuka, Masaru, 26, 192n
Ideation, 88ff, 163
Ideation
  classical view of, 88
  knowledge level model of, 102ff, 115
  process of, 91, 149
  variation-selection theory of, 92ff, 98, 99
  Wallas model of, 88ff
Ill-structuredness, 43, 169
Illumination stage in ideation, 89, 90, 116
Incubation stage in ideation, 89, 90, 108
Industrial espionage, 63
Industrial Revolution, 24, 124, 134
Inference, 107, 108, 109
Inference rules, 39
Information processing paradigm, 7, 102
Instantiation, 107, 109
Intellectual gap between scientists and artisans, 207n
Intentionality, 27, 31, 181
Intermetallic compound, 200n
Introspection, 89, 95, 202n
Inventing, characteristics of, 183ff
Invention, 50, 52, 53ff, 64
Inventionhood, test of, 55ff, 64
Ion of Ephesus, xiii
Iron-carbon diagram, 46
Iron-making, 21
Ishikawa, Masaru, 26, 192n

James, C. L. R., 8, 188n
James, William, 95
Jeffrey, Linda, 90, 202n
Johnson-Laird, Philip, 193n, 214n
Jones, Christopher, 5, 12, 178, 187n, 188n, 214n
Jordan, Michael, 46
Josephson, S. G., 193n, 195n
Jurgen, R. K., 192

Kamminga, Johan, 174, 213n
Kanigel, Robert, 57, 60, 197n, 198n
Kant, Elaine, 74, 75, 99, 117, 200n, 204n
Kelly, William, 62
Kepler, Johannes, 8, 37, 194n
Kepler's laws of planetary motion, 33, 56, 186
Keynes, John Maynard, 198n
Kilburn, Tom, 104, 162, 191n
Knowledge
  compiled, 51, 121
  declarative, 33
  deep, 51, 174
  failure as a source of, 173, 174
  heuristic, 34

immediate, 120
incomplete, 43
network of, 45
nonverbal, 178
operational, 37
in structural engineering, 18, 156, 160
types of, 32ff
Knowledge body, 33, 39, 46, 48, 56, 119
Knowledge level hypothesis on ideation, 102ff, 115, 150
Knowledge level model of cognition, 30, 31, 49, 56, 150, 181
  actions in, 38ff, 48
  goals in, 32ff, 48, 87
  knowledge in, 34ff, 87
Knowledge level process, 48ff, 56, 67, 87, 108
  design as a kind of, 49ff, 87
  nonrational, 49, 105
  rational, 48ff, 108
Knuth, Donald, 6, 26, 27, 37, 56, 166, 174, 187n, 192n, 194n, 197n, 212n, 213n
Koestler, Arthur, 49, 91, 101, 196n, 203n
Krebs, Hans, 103
Kubla Khan, 90
Kuhn, Thomas S., 7, 37, 52, 56, 84, 148, 166, 188n, 192n, 194n, 196n, 197n, 198n, 201n, 210n

Lachman, J. L., 188n
Lachman, R., 188n
Lahire, Phillipe, 175, 176
Laird, J., 193n
Lakatos, Imre, 201n
Lake, C. D., 210n
L'Allegro, 89
Langley, Patrick, 103, 204n, 210n
Larson, Judge Earl, 208n
Latency, 206n
Lattice parameter of a crystal, 200n
Laudan, Larry, 197n, 211n
Lavington, Simon, 191n
Lavoisier, Antoine, 3, 59
Layton, Edwin, 151, 152, 156, 187n, 211n
Leibniz, Gottlieb, 53, 55
Levesque, H., 195n
Linguistics, 31
Littlewood, J. E., 58
Loftus, E. F., 195n
Lowes, John Livingstone, 90, 202n

Maillart, Robert, 38, 179
Manchester Mark-I, 23, 54, 104, 191n
Manufacturing process
  as artifact, 14ff
  patenting of, 14, 15

Marginal solubility hypothesis, 71, 72
Marsh, J., xvii
Massachusetts Institute of Technology, 106, 144
Mathematical models in technology, 157
Mauchly, John, 21, 22, 23, 53, 54, 134, 138, 140, 141, 142, 144, 145, 147, 149, 162, 185, 191n, 208n, 210n
Mauchly, K. K., 208n
Mayr, Ernst, 203n
McClelland, J. L., 195n
McDermott, D., 62, 198n
McGannon, H. E., 190n, 196n
McNeil, Ian, 196n, 199n
McPherson, John, 144
Means-end heuristic, 115
Mechanics, 156
Medawar, Jean, 209n
Medawar, Peter, 17, 68, 190n, 200n, 209n
Memory, 14
Menai Strait, 18, 21, 28, 79, 111
Mental element, 93
Mental model, 178, 213n
Mental set, 47
Mental state, 27
Mentalism, 31
Metallurgical knowledge, 154, 155
Metallurgy, 4, 36, 46, 69, 175, 176
Metaphor, computation as, 31
Metropolis, N., 208n
Microprogrammed control unit, 23, 54, 103, 122, 183
Microprogramming, invention of, 23, 61, 103ff, 109, 122, 163, 183, 191n, 204n, 214n
Mill, John Stuart, 108, 205n
Miller, George A., 14, 188n
Milton, John, 89
Mind, 31, 38
Mining, 36
Minsky, Marvin, 193n, 196n
Modus ponens, 39, 111, 114, 205n
Mollenhoff, Clark, 163, 208n, 209n, 210n, 212n
Montrose Bridge, 112, 113
Moore School of Electrical Engineering, 22, 137
Moreland, Samuel, 128, 129, 131, 132
Mozart, Wolfgang Amedeus, 87, 202n
Multiple discovery in science, 92, 95
Mumford, Lewis, 62, 198n
Music, xiii

Nasmyth, James, 214n
Nature, xiii, xiv

Need, 4, 20ff, 26
Nervi, Pier Luigi, 38, 194n
Neural circuit, 30
Neurophysiology, 31
Newcomen engine, 54, 124ff, 131, 134, 137, 149, 160, 161, 183, 206n, 207n
Newcomen, Thomas, 54, 124, 131, 160, 206n
Newell, Allen, 29, 30, 31, 43, 74, 99, 102, 115, 117, 120, 167, 181, 192n, 193n, 195n, 200n, 204n, 205n, 212n
Newton, Isaac, 3, 37, 53, 55, 162
Newtonian mechanics, 59
Newtonian revolution, 59
Newton's rules of reasoning, 37
Nickel-titanium hardner, 72
Nimonic-75, 71, 87, 98
Nisbett, R. E., 195n, 205n
Node, 46
Nonmonotonic reasoning, 59, 197n
Normal design, 52, 84, 95
Normal science, 84
Normal technology, 95
Norman, Donald, 188n, 195n, 214n
North, Roger, 129
Novelty
   historical, 57, 61, 63
   psychological, 56
Nowak, G., 196n

Occam's razor, 37
Ontogeny, the biological concept, 209n
Ontogeny in technology, 123, 146, 148, 184
Operational principles
   definition, 158
   design and invention as source of, 170
   experience as source of, 171
   experiments as source of, 170, 171
   hypothesis, 167, 181
   interaction with theory, 159
   in the invention of computers, 162ff
   ontology of, 167ff
   as pictures, 177ff
   its relation to heuristics, 167
   its relation to reflection-in-action, 168ff
   science as source of, 174ff
   in software design, 164ff
   as technological knowledge, 157ff
   technological theory as source of, 174ff
Originality
   historical, 57, 58ff, 61, 63, 91, 95, 122, 204n
   its relation to creativity, 55ff
   psychological, 56, 58, 91, 95
Ornithine cycle, 103, 186

Pais, Abraham, 194n
Papert, Seymour, 193n
Papin, Denys, 126, 132, 133, 134, 161
Papin's piston-and-cylinder arrangement,
  161
Paradigm, 7, 31, 56, 66, 67, 155, 211n
Paradigm shift, 56, 59, 64
Parker, Jeanette, xvii
Pasteur, Louis, 20
Patrick, Catherine, 89, 90, 91, 202n, 203n
Peirce, C. S., 205n
Pencil, 11
Penrose, Roger, 193n
Perkins, David, 197n, 204n
Petroski, Henry, xvii, 11, 18, 43, 66, 123,
  148, 173, 188n, 190n, 195n, 199n,
  200n, 201n, 206n, 210n, 213n
Phlogiston theory, 33
Phylogeny, the biological concept, 209n
Phylogeny in technology, 123, 124, 132,
  146ff, 149, 160
Phylogeny law, 146, 148, 185
Picasso, Pablo, 8, 20
Pigou, A. C., 198n
Pitt-Rivers, Lane-Fox, 123
Planck, Max, 59, 196n
Planck's law, 47
Plastic design of structures, 25
Plastic design revolution, 60
Plato, xiii
Pneumatic process of steelmaking, 62
Poetry as divine dispensation, xiii
Poincaré, Henri, 88, 89, 90, 91, 202n, 203n
Polanyi, Michael, 157, 167, 168, 211n, 212n
Polya, George, 34, 194n
Popper, Karl, 12, 67, 87, 188n, 200n, 201n,
  202n
Positivism, 168ff
Powder metallurgy, 175
Precipitation hardening, 70, 71, 87, 98, 99,
  152, 200n
Prelude, The, 90
Preparation stage in Wallas' model, 89
Principle
  of bounded rationality, 45, 182, 195n,
    199n
  of faith, 38
  of parsimony, 37
  of rationality, 42, 43, 182
Principles of unity, 37
Printing by movable type, invention of, 59
Problem formulation, 105
Problem recognition, 105
Problem search, 116, 118, 120
Problem solving, 20, 23, 44, 200n

Problem space, 116, 117, 120
Propose-critique-modify strategy, 66
Proposition
  falsifiability of, 67, 82, 201n
  testability of, 67, 146
Protocol analysis, 74, 75, 76, 87, 99, 117,
  200n
Psychological novelty, 56
Psychological originality, 57
Puddling process, 15
Pye, David, 17, 50, 53, 55, 179, 187n, 189n,
  196n, 214n
Pylyshin, Zenon, 31, 192n, 193n

Qualitative law, 67
Quantum mechanics, 38
Quantum revolution, 59

Rabinow, Jacob, 26, 192n
Rajchman, Jan, 21, 144
Ramanujan, Srinivasan, 57ff, 197n
Ramsay, David, 127, 132
Randell, Brian, 148, 197n, 208n, 210n
Rationality, 42, 43, 66, 182, 199n
RCA Laboratories, 144
Reason, James, 18, 190n, 213n
Recapitulation law in biology, 209n
Reed-Hill, R. E., 200n
Reflection-in-action, 168
Reiter, Ray, 197n
Reitman, Walter, 43, 195n
Relativistic revolution, 59
Rendel, J. M., 80, 100, 112, 158
Rendel's trussing system, 112, 121, 158,
  159
Research tradition, 211n
Reverberatory furnace, 189n
Revolution
  in science, 52, 59, 103, 166, 195n, 198n
  in technology, 59, 60, 148, 166
Robison, John, 133, 207n
Rodin, Auguste, xiii
Rolling mill, 46
Rolf, L. T. C., 129, 131, 133, 161, 189n,
  190n, 206n, 211n
Rosenberg, Nathan, 63, 64, 159, 160, 177,
  188n, 190n, 191n, 199n, 200n, 201n,
  211n, 213n
Rosenbloom, P. S., 193n
Rota, G-C., 208n
Royal Society, 3, 129, 133, 161
Rules, 33, 39
Rumelhart, David, 195n
Ruse, Michael, 203n
Ruskin, John, 13

Santayana, George, 173
Savery, Thomas, 129, 131, 132, 133, 161
Savery's engine, 129, 130, 132, 161
Scale model, 43, 84, 171
Scaling up, 43
Shank, Roger, 196n
Schelp, Helmut, 148
Schemata, 47, 50
Schon, Donald, 168, 169, 170, 212n
Science, xii, xiv, 3, 6, 8, 17, 25, 92
Science as the understanding of nature, xiii
    paradigms in, 7, 31, 56, 66, 67, 155
    revolutions in, 52, 103
Sciences, natural, 4, 37, 58, 151, 154, 155
    engineering, 4, 59
Search
    in problem solving, 20, 84
    its role in ideation, 116ff
Search space, 98
Searle, John, 192n, 193n
Semantic network, 195n
Semiconductor memory, 16
Shannon, Claude, 163, 212n
Shaw, Cliff, 102
Shaw, Mary, 166, 211n, 212n
Shear steel, 62, 192n
Sheffield, 46, 63
Shiffrin, R. M., 188n
Shrager, J., 210n
Silverstein, M., 194n
Simon, Herbert A., 4, 5, 10, 29, 43, 44, 102,
    103, 115, 167, 187n, 188n, 192n, 193n,
    195n, 199n, 200n, 204n, 205n, 212n
Simonton, Dean Keith, 92, 93, 94, 96, 100,
    203n
Simonton's model of ideation, 93ff
Sintering, 175, 177
Smelting, 21, 36
Smiles, Samuel, 15, 21, 62, 186, 189n, 190n,
    192n, 198n, 199n
Smith, Cyril Stanley, 156, 211n, 212n
Smith, John Maynard, 203n
Smithers, Tim, xvii
Snow, C. P., 57, 58, 197n
Socrates, xiii
Software
    as artifact, 15ff, 36, 164, 165, 189n
    design, 165, 166
    engineering, 16, 36, 166
    Janus-like nature of, 15ff
    mathematical theory of, 165, 166, 189n
Somerscales, E. F. C., 197n
Sommerville, Ian, 189n
Sperry, Roger, 38, 194n
Spreading activation, 45, 48, 108, 181, 195n
Steadman, Philip, 123, 206n

Steam
    engine, 54, 161, 162, 183, 206n
    pressure, 127, 207n
Stearns, Joseph, 24
Steel, 46, 47
Steel-frame structures, 25, 60
Steel making, 24, 53, 55, 62, 63, 121, 192n
Steel Structures Research Committee (U.K.),
    25
Stephenson, George, 79
Stephenson, Robert, 17, 78, 79, 80, 81, 101,
    110, 111, 121, 122, 157, 158, 183,
    201n, 204n, 205n, 211n
Sternberg, Robert, 204n
Stored program digital computer, 15, 21, 53,
    54, 59
Sturt, George, 9, 13, 165, 172, 188n, 212n,
    213n
Strength of materials, theory of, 156, 187n,
    201n
Structural theory, 155, 157, 165, 187n
Structured programming, 36, 165, 166
Structured set of actions, 42, 48, 108
Structures, design of, 25, 50, 66, 165
Sullivan, Louis, 46
Superalloys, 21, 69ff, 97ff, 149
Symbol level of cognition, 30
Symbol processing paradigm, 7
Symbol structures, 12, 31, 40, 41, 48, 51

Tacit knowledge, 212n
Talking sketch, 179
Technical rationality, 168ff
Technological creativity, xiii, xiv, xv, xvii,
    30, 49, 90, 93, 95, 141
    a theory of, 7, 180ff
Technological ideas, xiv
Technological knowledge, 34, 92, 150ff,
    156ff, 167, 181, 194n, 200n, 211n
Technological problems
    role of curiosity in, 25ff, 181
    role of dissatisfaction in, 20, 22ff, 26, 27,
        137, 141, 181
    role of goals in, 27, 28
    role of need in, 20ff, 26, 181
Technological reasoning, 66, 87, 151, 152
Technological theory, 152, 155, 158, 171,
    174
Technological worldview, 37, 150
Technologist
    as a cognitive being, 28, 37, 108, 180ff
    as intellectual, 3, 5
Technology
    as a cognitive process, 3, 7, 8
    as a creative process, xiv, 28
    definition of, 3, 9, 11

as an evolutionary process, 122ff, 148
as harnessing of nature, xiii, xiv
history of, xiii, 7, 14, 62, 162
as human enterprise, xiv
its logic, 3
as problem solving, 20, 27
its psychology, xiv, 3, 5
its relation to craft, 9, 12ff
its relation to science, xiii, 3, 6, 8, 17, 150, 154, 168ff, 187n, 211n
reasoning in, 8
role of history in, 122ff, 134
teleological nature of, 4
universal characteristics of, 6
Telegraph, 24, 27
Telephone, invention of, 24, 54, 55, 91, 103
Telford, Thomas, 79
Tension in beams, 160
TEX typesetting system, 26, 37, 166, 174, 192n
Thagard, Paul, 108, 195n, 198n, 204n, 205n
Theory of evolution, 53
Thinking sketch, 179
Thompson, E. P., 188n
Thornley, D. G., 187n
Thought
  experiment, 178
  product, 56
Throughput, 206n
Thurston, Robert, 127, 129, 131, 132, 207n
Timoshenko, Stephen, 63, 64, 175, 187n, 199n, 200n, 210n, 213n
Token, 32
Torricelli, Evangeliste, 126, 132, 161
Trial and error, 14, 34, 175
Turbojet engine, invention of, 148, 149
Turing, Alan, 31, 60, 61, 162, 193n, 198n
Turing machine, 60
Typography, 26, 27

Ultrasonic delay line memory, 21, 22
Unconscious thought, 89, 90, 91, 108
Uniform circular motion, 37
University Mathematical Laboratory, Cambridge, 23, 104
Usher, Abbott Payson, 127, 132, 206n, 207n

Vacuum, production of, 126
Vacuum tube, as circuit element, 141
Van Rijsbergen, C. J., xvii
Variation-selection theory, 92ff, 98, 99, 101, 104, 110, 116, 123
Verification stage in Wallas' model, 89ff

Vincenti, Walter, xvii, 63, 64, 84, 92, 95, 100, 154, 159, 160, 177, 188n, 190n, 191n, 196n, 199n, 200n, 201n, 202n, 203n, 204n, 211n, 213n
Vincenti's model of ideation, 95ff
Visual thinking, 179, 214n
Vitruvius, 14, 34, 43, 60, 174, 189n, 194n, 198n
von Guericke, Otto, 126, 131, 132
von Helmholtz, Hermann, 24, 88, 89, 91, 202n
von Ohain, Hans, 148
von Neumann, John, 23, 53, 134, 162, 196n

Wagner, Herbert, 148
Wagon wheel making, 172
Walker, Samuel, 63
Wallace, Alfred R., 53, 55, 59, 92
Wallas, Graham, 88, 108, 202n
Wallas' four-stage model of ideation, 89, 101
Wasteland, The, 29
Water, raising of, 126, 127
Watson, G. F., 192
Watson, James D., 55
Watt, James, 3, 124, 125, 133, 183, 214n
Watt's steam engine, 124, 183
Waves, The, 29
Weber, Robert J., xvii, 103, 197n, 204n
Weisberg, Robert, 88, 89, 90, 202n
Wheel, invention of, 59
Wheelwright's craft, 165, 172
Whirlwind computer, 106, 109, 121, 122, 164
Whitman, Walt, 13
Whittle, Frank, 148
Wilkes, Maurice V., 15, 23, 26, 27, 33, 42, 53, 61, 63, 91, 103, 121, 122, 148, 162, 163, 183, 189n, 191n, 192n, 193n, 197n, 198n, 205n, 210n, 214n
William of Occam, 37
William II, King, 129
Williams, F. C., 21, 54, 104, 162, 191n
Winston, Patrick, 193n
Wirth, Nicklaus, 36
Wolpert, Lewis, 187n
Woolf, Virginia, 29
Worcester, Marquis of, 127, 131, 132
Worcester's water-commanding engine, 127, 129, 130
Wordsworth, William, 90
Wright brothers, 59, 92, 202n
Wrought iron, 18, 64

X-ray diffraction, 72, 99, 153, 171

Zytkow, J. M., 204n